Introduction to GNSS Geodesy

Clement A. Ogaja

Introduction to GNSS Geodesy

Foundations of Precise Positioning Using Global Navigation Satellite Systems

Springer

Clement A. Ogaja
Boulder, CO, USA

ISBN 978-3-030-91823-1 ISBN 978-3-030-91821-7 (eBook)
https://doi.org/10.1007/978-3-030-91821-7

This Springer imprint is published by the registered company Springer Nature Switzerland AG
The registered company address is: Gewerbestrasse 11, 6330 Cham, Switzerland

To my parents, especially my mother

Preface

GNSS geodesy, or geodesy by means of global navigation satellite systems (GNSS), plays an important role in the accurate and precise locating of points and infrastructure anywhere on Earth. Take, for example, a modern application such as high-accuracy GNSS-supported augmented reality (AR) in the construction industry. Without accurate reference systems and precise GNSS geodetic positioning, it would be hard to produce accurate views of underground utilities that enable engineers in the field to see the true location(s) of buried pipes and conduits to avoid damage and safety issues. Even with the best of 3D software models, a low-accuracy GNSS data stream would cause the 3D model to move about on screen, making it impossible to confidently show where the virtual representation of an object should truly appear in the AR display's field of view. This is just one of the many examples of GNSS geodesy's role in modern-day applications.

This book introduces a selection of key elements of GNSS geodesy to help understand the concepts of achieving high precision from GNSS data, on the basis of the existing models and strategies. Discussions are provided with the assumption that the reader is primarily interested in understanding the key concepts and elements, and pertinent references are provided throughout as necessary for further reading and consultation.

Being an introductory book, it is intended as a complementary reference and guide for those involved in the fields of high-precision positioning with GNSS, including, but not limited to, students, researchers, scientists, engineers, and professionals in the fields of geodesy, geosciences, geomatics, engineering, geoinformatics, surveying, navigation, GNSS product and software development, and all related areas and applications.

While most of the illustrations were prepared by the author, some artwork and images have been used with permission from the copyright holders. The following individuals and institutions are acknowledged for copyright permissions to use their images and artwork: Augview Ltd (New Zealand), WestConnex (Sydney, Australia), Dr. Peter Steigenberger (DLR, Germany), Michael Bundock of Augview Ltd, and Stephanie Kirtland of Trimble Inc.; U.S. Department of Transportation (DOT), Federal Highway Administration (FHWA), The National Oceanic and Atmospheric

Administration (NOAA), U.S. Department of Commerce, International GNSS Service (IGS), and others are acknowledged for cited works and literature.

Thanks to the anonymous reviewers for their time spent reading the manuscript and providing valuable comments and suggestions. And most of all, to my family for their unconditional love and support.

Boulder, CO, USA
2021

Clement Ogaja

Contents

Acronyms

AFREF	African Reference Frame
AltBOC	Alternative Binary Offset Carrier
APC	Antenna Phase Center
APREF	Asia Pacific Reference Frame
AR	Augmented Reality
ARP	Antenna Reference Point
BDT	BeiDou Time
BeiDou	Chinese GNSS
BINEX	BINary EXchange
CDMA	Code Division Multiple Access
CEP	Celestial Ephemeris Pole
CMR	Compact Measurement Record
CoM	Center of Mass
CORS	Continuously Operating Reference Station
CTRF	China Terrestrial Reference Frame
DCB	Differential Code Bias
DD	Double Difference
DORIS	Doppler Orbitography and Radiopositioning Integrated by Satellite
ECEF	Earth Centred Earth Fixed
ENU	East-North-Up
ETRS89	European Terrestrial Reference System 1989
EUREF	European Reference Frame
FDMA	Frequency Division Multiple Access
FIG	Federation of International Surveyors
Galileo	European GNSS
GEO	Geostationary Orbit
GIS	Geographic Information Systems
GLONASS	Russian GNSS
GLNT	GLONASS Time
GNSS	Global Navigation Satellite System
GPS	Global Positioning System

GPST	GPS Time
GST	Galileo System Time
GTRF	Galileo Terrestrial Reference Frame
ICRF	International Celestial Reference Frame
ICSM	Australian and New Zealand Intergovernmental Committee on Surveying and Mapping
IERS	International Earth Rotation and Reference Systems Service
IFB	Inter-Frequency Bias
IGS	International GNSS Service
ISB	Inter-System Bias
ITRF	International Terrestrial Reference Frame
LAMBDA	Least-squares AMBiguity Decorrelation Adjustment
LCC	Lambert Conformal Conic
MEO	Medium Earth Orbit
MSM	Multiple Signal Messages
MW	Melbourne-Wubbena
NAD83	North American Datum 1983
NEQ	Normal Equation
NGS	National Geodetic Survey
NMEA	National Marine Electronics Association
NOAA	National Oceanic and Atmospheric Administration
NRCan	Natural Resources Canada
NSRS	U.S. National Spatial Reference System
NTRIP	Networked Transport of RTCM via Internet Protocol
PBO	Plate Boundary Observatory
PCO	Phase Center Offset
PCV	Phase Center Variation
PPP	Precise Point Positioning
PPP-RTK	PPP Real Time Kinematic
PR	Pseudorange
PRN	Pseudo Random Noise
PZ-90	Parametry Zemli 1990
QIF	Quasi-Ionosphere-Free
QZSS	Quasi-Zenith Satellite System
RCP	Right Circularly Polarized
RINEX	Receiver INdependent EXchange format
RTCM	Radio Technical Commission for Maritime Services
RTK	Real Time Kinematic
SBAS	Satellite Based Augmentation System
SD	Single Difference
SEGAL	Space and Earth Geodetic Analysis Laboratory
SIRGAS	Sistema de Referencia Geocéntrico para Las Américas
SLR	Satellite Laser Ranging
SPC	State Plane Coordinate
SSR	State Space Representation

STEC	Slant TEC
TAI	International Atomic Time
TD	Triple Difference
TEC	Total Electron Content
TECUs	TEC Units
UN	United Nations
UNOOSA	United Nations Office for Outer Space Affairs
UNRCC	United Nations Regional Cartographic Conference
US	United States
USGS	United States Geological Survey
UT1	Universal Time
UTC	Coordinated Universal Time
UTC(k)	Physical Realisation of UTC
UTM	Universal Transverse Mercator
VCV	Variance-Covariance
VLBI	Very Long Baseline Interferometry
WGS-84	World Geodetic System 1984

The original version of the book has been revised. A correction to this book can be found at https://doi.org/10.1007/978-3-030-91821-7_8

Part I
Introduction

This book starts with a use case example to illustrate how geodetic GNSS positioning is a boon to some of the modern practical high-tech applications. By focusing on just one example, it is shown how GNSS geodesy supports high-accuracy Augmented Reality for various applications, for example, in the construction industry, utilities, natural resources, land development, and other areas. The author hopes that by starting off with a practical geodetic GNSS use case example, the reader is inspired to learn, in Part II, the key elements of GNSS geodesy that makes accurate and precise geopositioning possible. For example, it is important to understand the geodetic reference systems and reference frames and the associated GNSS data processing strategies that enable both accurate and high-precision geopositioning.

Chapter 1
Augmented Reality: A GNSS Use Case

1.1 Accurate Geopositioning

An Augmented Reality (AR) technology overlays digital information such as a virtual 3D model (of a building design, a road design, buried pipes, utility cables, or any real objects) on a live image of what is being viewed through a device such as a smartphone camera. The intent of AR is to overlay the virtual 3D model either as a copy of the real object as it exists in situ or as a design of something to be added or built, at the location being viewed through the camera display screen of the device. Primarily, an augmented reality system comprises a mobile device, handheld or head-mounted, with an integrated display screen, a camera, processor, a global positioning system, inertial orientation sensing, microphone, and so forth. Since modern smartphones already have such hardware, AR systems have been developed that makes use of the same as shown in Fig. 1.1. Many studies have looked at AR applications in the fields of agriculture, architecture, building construction, civil construction and inspection, surveying and mapping, residential development, urban transportation and planning, underground utility inspections, and other areas (see, e.g., [2, 3, 5, 9, 11]).

In cases where the real object of interest is invisible through the live image (e.g., the case of buried underground utility cables), it is even more critically important that the 3D model being viewed live through the camera display screen depicts the actual "true" location of the object. In such situations, a machine operator is in danger of damaging underground cables and causing safety issues for themselves if the AR system displays incorrect location of the underground cables. This is where GNSS geodesy, or precise positioning with Global Navigation Satellite Systems, comes in.

By NOAA's definition, "*Geodesy is the science of accurately measuring and understanding the Earth's geometric shape, orientation in space, and gravity field. ... Geodesists must accurately define the coordinates of points on the surface of the*

C. A. Ogaja, *Introduction to GNSS Geodesy*,
https://doi.org/10.1007/978-3-030-91821-7_1

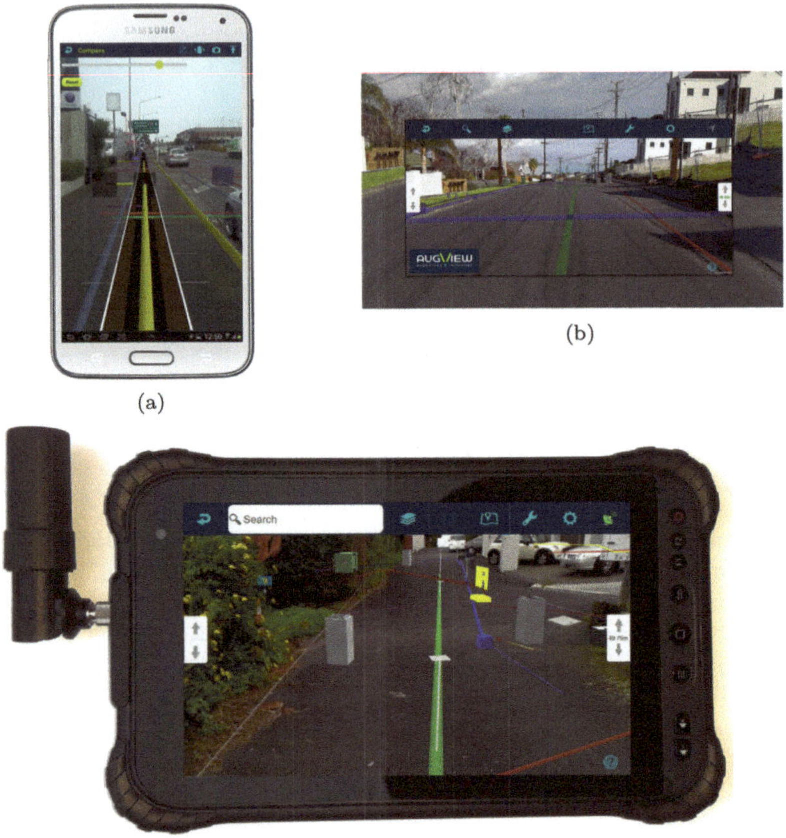

(a) (b) (c)

Fig. 1.1 Screenshots of augmented reality (AR) display systems. (**a**) AR on smartphone. (**b**) An AR of a street scene. (**c**) AR on ruggedized RTK tablet. Images used by permission. Courtesy of Augview Ltd

Earth in a consistent manner." Therefore, GNSS geodesy enables accurate location of points and the shape and size of the Earth by means of Global Navigation Satellite Systems.

The AR system (Fig. 1.1) uses a global positioning system (GPS) or GNSS among other components. This provides the position of the antenna from which the placement of the virtual image in the live field of view is defined. However, even with the best of 3D modeling software, *a low-accuracy GNSS data stream would cause the 3D model to wander on the AR screen, making it impossible to confidently show where the virtual representation of an object should truly appear in the display's field of view*.

The use of external geodetic GNSS antenna/receiver unit as shown in Fig. 1.2 can provide accurate GNSS data streams to the AR system. However, even with

Fig. 1.2 A geodetic GNSS-based augmented reality system comprised of a geodetic antenna+receiver, data communication cable, and a mobile device with a display, camera, inertial orientation sensors, and other components. Image used by permission. Courtesy of Trimble Inc

the most accurate data streams, it would still be impossible to correctly, "virtually," locate objects in situ without an accurate geodetic datum of reference (for correctly and accurately displaying the locations of real-world objects or points). Such datums are only established through the science and practice of GNSS geodesy.

Figure 1.3 illustrates how a high-accuracy AR system is supported by GNSS geodesy. Layer A (geodetic infrastructure by governments and international organizations for the science and definition of an accurate reference system, as well as other purposes [6, 7]) supports Layer B (GNSS data stream from service providers, organizations, and GNSS professionals), which subsequently supports the end user (positioning of the AR system). The geodetic infrastructure includes physically established reference stations and tracking networks. The accurate precise positioning of the AR system antenna is provided through real-time kinematic (RTK) GNSS data stream which relies on the accurate definition of a reference system (aka datum). Chapter 3 is dedicated to the topic of reference systems in GNSS geodesy. This is complemented in Chap. 6, which discusses some of the existing GNSS geodetic infrastructure for actualizing the reference systems.

Smartphone's in-built GNSS chips are nowadays capable of precise carrier phase measurements, from multiple constellations and frequencies. Therefore, the users may not always need to rely on dedicated external units as shown in Fig. 1.2. Real-time delivery of products such as RTK corrections, precise orbits and clocks, atmospheric corrections, and satellite hardware biases for precise point positioning

Layer A: GNSS Geodetic Infrastructure

Networks of GNSS continuously operating reference stations (CORSs); continuously archived raw and processed data; geodetic products (for example, CORS coordinates and velocities in a national or global reference system; satellite orbits; clocks; atmospheric corrections; biases); and web-based interfaces for offering users access to a reference system, e.g., by tying surveyed data to the reference system (georeferencing).

Layer B: GNSS Service Providers

GNSS CORS (base station) service, for example, from GNSS receiver manufacturers and product developers. Service owner, station custodian sets the position coordinates for base station by using a Layer A service such as NOAA's Online Positioning User Service (OPUS). For instance, at the time of establishing or updating the station, collects long duration data (2 hours or longer), submits to OPUS and uses results as the known fixed coordinates for the station. The coordinates will be known in a specified reference system (from Layer A).

Real time delivery of products such as RTK corrections in RTCM formats via NTRIP; real time delivery of precise orbit and clocks, atmospheric corrections, and satellite hardware biases for PPP method.

Layer C: GNSS Receiver in AR system

Computes real-time geodetic position coordinates of the AR system antenna by using data stream from Layer B. The antenna coordinates computed in real-time are in a reference system established by geodetic infrastructure (Layer A), and are thus georeferenced.

Fig. 1.3 Service layers illustrating how an augmented reality (AR) system is supported by GNSS geodesy. Layer A (GNSS geodetic infrastructure by governments and international organizations) supports Layer B (GNSS data stream from service providers, organizations, or GNSS professionals), which subsequently supports the end user (the AR system precise positioning)

(PPP) thus becomes important. It should be noted though that GNSS geodesy is the source of such products (i.e., they are generated from GNSS geodetic data processing).

1.2 High-Accuracy AR Applications

AR technology supported with increased location accuracy allows 3D design and construction models to be virtually overlaid in their real-world position, in a user's view of existing environment. The AR display of a design of what is not yet constructed enables users to see the intent in a real-world context and provides the opportunity to compare design alternatives in context, check relationships between existing and future elements, and monitor site logistics and equipment movements. And after construction, AR can overlay and compare 3D design models onto the end result in the field to inspect the construction and monitor compliance with the original design intent.

AR devices can provide a 3D virtual view of existing elements that are not visible to the user in the real-world environment, such as buried underground utilities (cables, gas pipes, and drainage pipes) or structural components that are hidden from the current user's view of the structure. This helps to verify that existing 3D models of underground utilities are accurate, to visualize potential conflicts between the proposed design and existing facilities, and to communicate any issues to affected project stakeholders.

Some of the examples of AR applications in the real-world construction and utilities projects to date are described in [1] and [4]. These include AR use in projects by Utah Department of Transportation (UDOT), Florida Department of Transportation (FDOT), Michigan Department of Transportation (MDOT), and a general contractor in Sacramento, California. Figure 1.4 is a sample snapshot of an AR use in Michigan.

The M4-M5 Link Tunnels Project in Sydney, Australia, is another example of a construction project that has taken advantage of the latest AR technology by using a product known as SiteVision [10]. The SiteVision is a smart device application that connects to satellite GNSS and overlays the design data on top of the work location in real time, providing information on building features and underground services such as conduits, pits, and pipes (Figs. 1.5, 1.6, and 1.7). Its geodetic component ensures the GPS/GNSS accuracy improves from five meters (16 ft) down to two centimeters (a 0.8 in.).

Figure 1.8a, from a case study in the UK, shows a road upgrade project in which a proposed new bridge is displayed at its on-site location. The project used Trimble's

Fig. 1.4 A snapshot of AR use in Michigan to propose design elements, overlaid onto the existing conditions on the bridge carrying Milwaukee Avenue over I-75. Adapted from [1]

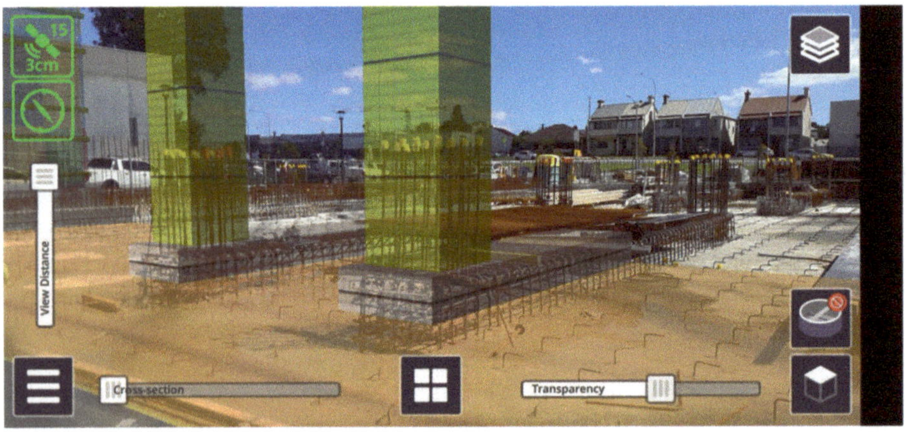

Fig. 1.5 GNSS-based augmented reality in The M4-M5 Link Tunnels project in Sydney, Australia, in 2021. The smart app known as SiteVision connects to satellite GPS and overlays the road design data on top of the work location in real time. Image used by permission. Courtesy of WestConnex (westconnex.com.au)

SiteVision[1] to overlay 3D models from CAD and engineering software in live real-world views.

[1] Running on an android-based smartphone, SiteVision uses Trimble Catalyst technology and precise GNSS to determine its position accurate to 1 to 2 centimeters. SiteVision then overlays a 3D model onto a real-time image of the site [10].

Fig. 1.6 A snapshot of SiteVision AR in real-time overlay of the design data of pipes and conduits onto the existing conditions at the work location in The M4-M5 Link Tunnels project in Sydney, Australia, in 2021. Image used by permission. Courtesy of WestConnex (westconnex.com.au)

Fig. 1.7 A snapshot of SiteVision AR in real-time overlay of design data of the tunnel facade onto the existing conditions at the work location in The M4-M5 Link Tunnels project in Sydney, Australia, in 2021. Image used by permission. Courtesy of WestConnex (westconnex.com.au)

Figure 1.8b is an example of SiteVision being used to locate underground assets and utilities in a built-up environment. Using precise AR to indicate the location of underground assets such as utility cables and pipelines is an important safety consideration in any project that involves digging up of the ground and/or putting up of new structures.

Figure 1.8c is an example of AR application to oil and gas pipelines [8], in which the AR app provides real-time on-site views of existing elements or proposed

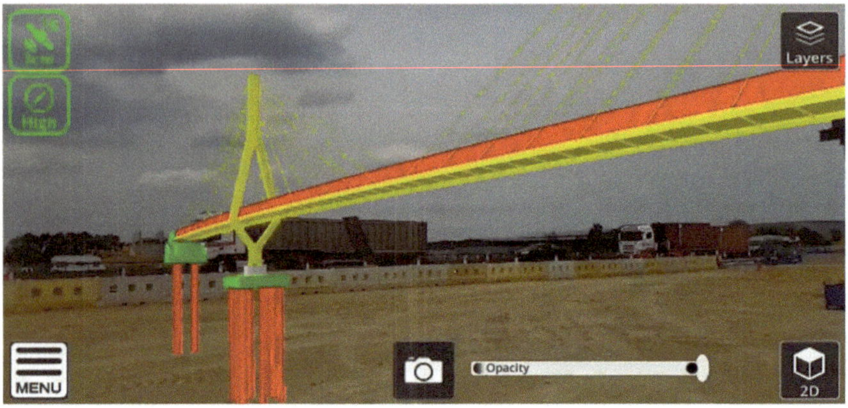

(a)

(b)

(c)

Fig. 1.8 High-accuracy GNSS-based augmented reality (AR) applications. (**a**) GNSS-based augmented reality (AR) in civil engineering. (**b**) GNSS-based augmented reality (AR) in underground asset location. (**c**) GNSS-based augmented reality (AR) in oil and gas industry. Images used by permission. Courtesy of Trimble Inc

improvements, and tools such as SiteVision can be used for planning oilfield work such as pump stations and well heads.

In developing natural resources such as oil and gas, contractors often need access to land they do not own. The process typically involves negotiation with landowners, for instance, to decide the location and terms of an easement or right-of-way. This would require visits to the affected property sites with 2D maps and detailed written descriptions which landowners may find hard to interpret or visualize in terms of how their properties would be affected. In such cases, it is beneficial to use AR (instead of 2D paper drawings and descriptions) to provide accurate, on-site visualizations of proposed easements or structures when working with landowners.

Other applications include, for example, use of AR to evaluate a planned land development or residential subdivision project before the construction begins. By using a real-time on-site visualization of the planned project, the AR display can help survey crews to verify stakeout to ensure that designs fit the planned site and are staked correctly.

1.3 Concluding Remarks

There are many application areas that benefit from the high-accuracy precise positioning by means of Global Navigation Satellite Systems (GNSSs). The augmented reality (AR) technology is one such area especially as is currently being used in architecture, construction, surveying, mapping, transportation, utilities, and other outdoor activities. As reviewed in this chapter, the AR concept, while comprised of different technology components, is dependent on GNSS for accurate and precise geoposition information. The examples of AR applications in construction and other areas serve to demonstrate how high-accuracy augmented reality is increasingly getting attention by both product developers, or innovators, and users of the technology in outdoor field projects and megaprojects. High-accuracy virtual location of both visible and invisible elements is achievable due to advances in the theory and practice of GNSS Geodesy, which supports product development and provision of pertinent GNSS positioning infrastructure. The subsequent chapters of this book are devoted to understanding the theory of GNSS constellations, signals, and the key elements of GNSS geodesy.

References

1. Azari H., & Gilson K. (2020). Can augmented reality address highway construction challenges? Publication Number: FHWA-HRT-20-004, vol. 84, no. 2. U.S. Department of Transportation, Federal Highway Administration.
2. Bae, H., Golparvar-Fard, M., & White, J. (2013). High-precision vision-based mobile augmented reality system for context-aware architectural, engineering, construction and facility management (AEC/FM) applications. *Visualization in Engineering, 1*, 3. https://doi.org/10.1186/2213-7459-1-3

3. Gazcón N., Nagel J., Bjerg E., & Castro S. (2018). Fieldwork in Geosciences assisted by ARGeo: A mobile augmented reality system. *Computers & Geosciences, 121*, 30–38. ISSN 0098-3004. https://doi.org/10.1016/j.cageo.2018.09.004
4. Gilson K., Mallela J., & Goodrum P. (2020). Leveraging augmented reality for highway construction. Publication Number: FHWA-HRT-20-037. U.S. Department of Transportation, Federal Highway Administration.
5. Liu F., & Seipel S. (2018). Precision study on augmented reality-based visual guidance for facility management tasks. *Automation in Construction, 90*, 79–90. ISSN 0926-5805. https://doi.org/10.1016/j.autcon.2018.02.020
6. National Academies of Sciences, Engineering, and Medicine (2020). *Evolving the geodetic infrastructure to meet new scientific needs*. Washington: The National Academies Press. https://doi.org/10.17226/25579
7. National Research Council (2010). *Precise geodetic infrastructure: National requirements for a shared resource*. Washington, DC: The National Academies Press. https://doi.org/10.17226/12954
8. Radopoulou, S. (2020). Putting AR to work in the field. *World Pipelines North America, 20*(12), 26–28. ISSN 1472-7390.
9. Schall, G., Mendez, E., Kruijff, E., Veas, E. E., Junghanns, S., Reitinger, B., & Schmalstieg, D. (2009). Handheld Augmented Reality for underground infrastructure visualization. *Personal and Ubiquitous Computing, 13*, 281–291. https://doi.org/10.1007/s00779-008-0204-5
10. Trimble Inc. (2021). Trimble Sitevision—High accuracy augmented reality system. Retrieved May 25, 2021 from https://sitevision.trimble.com/
11. Zheng, M., & Campbell, A. (2019). Location-based augmented reality in-situ visualization applied for agricultural fieldwork navigation. In *2019 IEEE International Symposium on Mixed and Augmented Reality Adjunct (ISMAR-Adjunct)*. https://doi.org/10.1109/ISMAR-Adjunct.2019.00039

Chapter 2
GNSS Constellations and Signals

2.1 GNSS Constellations

A Global Navigation Satellite System, GNSS, is comprised of three components: *space segment*, *control segment*, and *user segment*. Space segment is the satellites orbiting the Earth; control segment is the infrastructure monitoring and operating the satellites; and the user segment is all users of the GNSS signals being broadcast by the space-borne satellites. Figure 2.1 illustrates how these individual components interact to enable users to calculate their position anywhere on the Earth or in space.

The space segment, also referred to as orbit *constellation*, is a network of space-borne satellites orbiting the Earth in equally spaced predictable orbits at a specified altitude. All GNSS orbits are MEOs (Medium Earth Orbits). In comparison, geostationary satellites orbit the Earth at much higher altitudes, about twice that of GNSS satellites. At a minimum, each of the GNSS orbit constellations (Fig. 2.2) were designed so that at least four satellites are visible anywhere on the Earth at any time. The individual orbit design characteristics are as shown in Table 2.1.

Each GNSS satellite broadcasts radio signals (Sect. 2.2) that receivers can use to calculate a position. Information on satellite health, satellite position, as well as data that can be used to determine the satellite time are all transmitted via these signals. See additional notes in Sect. 2.2.

The integrity of a constellation relies on the satellites precisely maintaining their orbits, and it is imperative that the satellite positions in space be monitored. This is a role of the control segment. As an illustration, in the case of the GPS constellation, the monitor stations (Fig. 2.1) track all satellites in their range and collect data of the satellite signals. The raw data are sent to the master control station where they are processed, and new information about orbits, ephemeris, and clocks of the satellites are calculated and uploaded to each satellite on a regular basis. The uploaded information provides a basis for almanac and ephemeris for the user segment (*almanac* are the course orbital parameters for all satellites; *ephemeris* data are precise orbital and clock corrections for each satellite).

C. A. Ogaja, *Introduction to GNSS Geodesy*,
https://doi.org/10.1007/978-3-030-91821-7_2

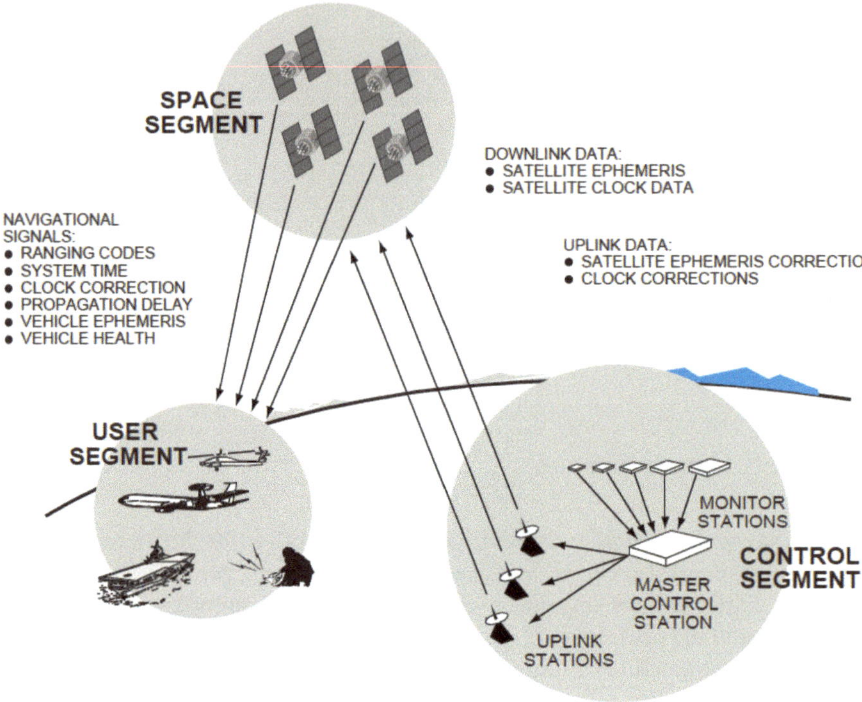

Fig. 2.1 Three components of a GNSS system

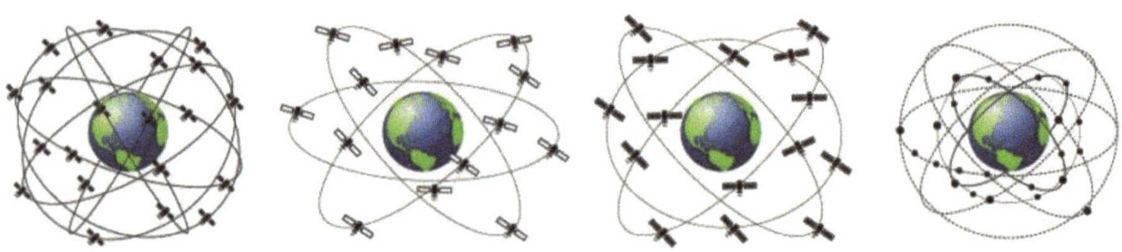

Fig. 2.2 GPS, Galileo, GLONASS, and BeiDou constellations, from left to right in that order. Orbital characteristics are shown in Table 2.1

The orbital data are uploaded to the satellite in a compact format, e.g., using Keplerian elements (see Appendix A). This simplifies the process of predicting the satellite orbits. For instance, given time information and the Keplerian elements for any satellite, its time-dependent positions can be predicted in X-Y-Z coordinate format (e.g., [7]). Knowing these time-dependent X-Y-Z positions of every satellite is an important part of calculating a position for the user located anywhere on the Earth or in space.

Table 2.1 GNSS constellation characteristics

System	Altitude (height above sea level)	Orbital period	Other characteristics
GPS	20,200 km (12,550 mi)	11 h 58 min	Global coverage; 6 orbital planes; 55° inclination angle
Galileo	23,222 km (14,430 mi)	14 h ** min	Global coverage; 3 orbital planes; 56° inclination angle
GLONASS	19,100 km (11,900 mi)	11 h 15 min	Global coverage; 3 orbital planes; 64.8° inclination angle
BeiDou	38,300 km and 21,500 km (various)	—	Global coverage; 6 orbital planes; 55° inclination angle

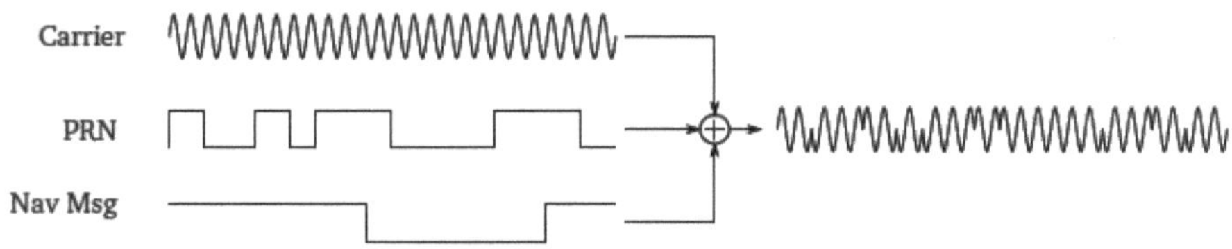

Fig. 2.3 Signal structure of a GNSS carrier wave modulated with PRN ranging code and navigation message

2.2 GNSS Signals

Satellites broadcast radio signals to enable GNSS receivers to determine location and time. The GNSS signals include ranging signals, used to measure the distance to the satellite, and navigation messages. The navigation messages include ephemeris data, used to calculate the position of the satellite in orbit, and information about the time and status of the satellite constellation. For the ranging signals and navigation message to travel from the satellite to the receiver, they are modulated onto a carrier wave for signal transmission. This is done by means of phase modulations (Fig. 2.3).

Positioning with GNSS satellites is based on the concept of trilateration, a process in which the user position is calculated from distances (ranges) to multiple GNSS satellites in combination with the known positions of those satellites. The basic principle of measuring a satellite-receiver distance (range) involves knowing the time required for a signal to propagate from the satellite to the user receiver. Estimating the satellite-receiver distance can be based on one of the two approaches, namely: (i) code phase-based *pseudorange* determination using time-shift from

PRN[1] code correlation, and (ii) *carrier phase*-based range determination using the satellite's carrier signal instead of the coded messages contained within the carrier signal.

The GNSS signals are generated onboard the GNSS satellites and are based on the frequency output and the time signal of the satellite's atomic clock. The atomic clock provides the nominal frequency used to generate the different carrier frequencies of the GNSS signals and the satellite time signal used to modulate the carrier wave to enable measurements of the satellite time at the receiver and to transmit information to the user.

Table 2.2 shows the nominal carrier frequencies for GPS, Galileo, GLONASS, and BeiDou constellations [1, 5, 6, 8–11]. The Galileo signals on frequencies E5a and E5b are combined by the so-called AltBOC[2] modulation, generating a wideband signal with the effective carrier frequency of 1191.795 MHz (a frequency band also named as E5).

GPS, Galileo, and BeiDou constellations use a technique known as Code Division Multiple Access (CDMA)[3] to transmit navigation signals from multiple satellites on same nominal frequency. The CDMA method is used to distinguish

Table 2.2 GPS, Galileo, GLONASS, and BeiDou signals

System	Band (component)	Frequency MHz	Wavelength cm
GPS	L1	1575.420	19.03
	L2	1227.600	24.42
	L5	1176.450	25.48
Galileo	E1	1575.420	19.03
	E5a	1176.450	25.48
	E5b	1207.140	24.83
	AltBOC	1191.795	25.15
	E6	1278.750	23.44
GLONASS[a]	G1	1602+k×9/16	**.**
	G1a	1600.995	18.73
	G2	1246+k×7/16	**.**
	G2a	1248.060	24.02
	G3	1202.025	24.94
BeiDou	B1	1561.098	19.20
	B2	1207.140	24.83
	B3	1268.520	23.63

[a] For FDMA-based GLONASS satellites, the actual frequency of signal transmission on G1 and G2 is derived from the channel number $k = -7, \ldots, 6$

[1] Pseudo-Random Noise.

[2] Alternative Binary Offset Carrier (AltBOC).

[3] CDMA is a method of frequency reuse whereby multiple radios use the same frequency channel but each one has a unique code.

between satellites when transmitting navigation signals. For instance, GPS satellites all transmit on the same nominal frequency (L1/L2/L5), but with different PRN codes. A GPS receiver gets a unique segment of the C/A and P codes from each satellite but all on a single carrier frequency on which they are transmitted. Therefore, even though each satellite broadcasts its own unique segment of the C/A code, or PRN, they all arrive at the receiver at the same frequency. The same is true of the P code, each unique segment of the P code (a satellite's PRN) arrives at the same frequency. For each of the nominal frequencies shown in Table 2.2, CDMA-based satellites all use same frequency channel, but each satellite has a unique PRN code. GLONASS mostly uses Frequency Division Multiple Access (FDMA) to distinguish between satellites whereby each satellite transmits its navigation signals on a slightly different frequency. However, as of this writing, there are some newer GLONASS satellites that use the CDMA method [14].

Due to multiple constellations and frequencies, GNSS signals are not only based on the carrier frequencies but also on how the navigation information is modulated onto the carrier and provided to the users [13]. The RINEX (Receiver INdepndent Exchange) format, discussed in Sect. 6.3.1 of Chap. 6, is an example of how multi-frequency GNSS data recorded by receivers can be decoded for easy exchange and processing by various independent GNSS processing software. RINEX was proposed and developed on the understanding that most GNSS processing software use a set of well-defined observables that include the carrier phase measurements at one or multiple frequencies, pseudorange measurement, and observation time.

The RINEX format explains the multiple options available to a user when processing multi-frequency GNSS data. It identifies individual GNSS signals based on unique three character IDs (aka. observation codes) assigned by a receiver to individual satellites according to the frequency band and tracking channel or code. Tables 2.3, 2.4, 2.5, and 2.6, respectively, show examples of the mapping of GPS, Galileo, GLONASS, and BeiDou satellite signals to both carrier phase and pseudorange observation codes.

Precise positioning techniques rely on measuring the phase of the carrier wave on which the GNSS signals are modulated. The phase observable[4] has a much lower noise level compared to the pseudorange observable,[5] and thus the use of phase

Table 2.3 Examples of GPS observation codes [4]

Band (freq.)	Code	Carrier phase	Pseudorange
L1	C/A	L1C	C1C
	P (AS off)	L1P	C1P
L2	C/A	L2C	C2C
	P (AS off)	L2P	C2P
L5	Q	L5Q	C5Q
	I+Q	L5X	C5X

[4] Carrier phase observation codes in Tables 2.3, 2.4, 2.5, and 2.6.

[5] Pseudorange observation codes in Tables 2.3, 2.4, 2.5, and 2.6.

Table 2.4 Examples of Galileo observation codes [4]

Band (freq.)	Code	Carrier phase	Pseudorange
E1	A PRS	L1A	C1A
	C OS Pilot	L1C	C1C
E5a	Q no data	L5Q	C5Q
	I+Q	L5X	C5X
E5	Q	L8Q	C8Q
	I+Q	L8X	C8X

Table 2.5 Examples of GLONASS observation codes [4]

Band (freq.)	Code	Carrier phase	Pseudorange
G1	C/A	L1C	C1C
	P	L1P	C1P
G2	C/A	L2C	C2C
	P	L2P	C2P
G3	Q	L3Q	C3Q
	I+Q	L3X	C3X

Table 2.6 Examples of BeiDou observation codes [4]

Band (freq.)	Code	Carrier phase	Pseudorange
B1	Q	L2Q	C2Q
	I+Q	L2X	C2X
B2	Q	L7Q	C7Q
	I+Q	L7X	C7X
B3	Q	L6Q	C6Q
	I+Q	L6X	C6X

observable leads to a higher positioning accuracy. However, the phase observable is ambiguous by an unknown number of wavelengths and has to be resolved in the positioning process. The process for achieving ambiguity resolution is presented later in Chap. 7.

The modeling of GNSS observables to reduce or eliminate error sources affecting precise positioning is an important part of GNSS data processing. Such error sources are reviewed in Chap. 5. As part of the error sources, observables are biased by delays induced by receiver and satellite hardware. These are briefly reviewed in Sect. 5.3.2.5 of Chap. 5, and further details can be found, e.g., in [2, 3, 12] and other sources.

References

1. GLONASS ICD. (2008). *GLONASS Interface Control Document L1/L2* (5.1 ed.). Russian Institute of Space Device Engineering.
2. Håkansson, M., Jensen, A. B. O., Horemuz, M., et al. (2017). Review of code and phase biases in multi-GNSS positioning. *GPS Solution 21*, 849–860.

3. Håkansson, M. (2020). *GNSS Hardware Biases in Code and Carrier Phase Observables*. KTH School of Architecture and The Built Environment, Stockholm, Sweden. ISBN 978-91-7873-454-2.
4. IGS. (2020). *RINEX: The Receiver Independent Exchange Format, Version 3.05*. International GNSS Service (IGS), RINEX Working Group and Radio Technical Commission for Maritime Services Special Committee 104 (RTCM-SC104). https://files.igs.org/pub/data/format/rinex305.pdf
5. IS-GPS-200. (2021). *GPS Interface Specification IS-GPS-200, Revision M*. Global Positioning Systems Directorate.
6. IS-GPS-705. (2021). *GPS Interface Specification IS-GPS-705, Revision H*. Global Positioning Systems Directorate.
7. Ogaja, C. (2011). *Applied GPS for engineers and project managers*. Reston: ASCE Press. https://doi.org/10.1061/9780784411506
8. OS-B1I-SIS-ICD. (2019). *BeiDou Open Service B1I Signal in Space Interface Control Document, Version 3.0*. China Satellite Navigation Office.
9. OS-B3I-SIS-ICD. (2018). *BeiDou Open Service B3I Signal in Space Interface Control Document, Version 1.0*. China Satellite Navigation Office.
10. OS-B2a-SIS-ICD. (2017). *BeiDou Open Service B2a Signal in Space Interface Control Document, Version 1.0*. China Satellite Navigation Office.
11. OS-SIS-ICD-2.0. (2021). *Galileo Open Service Signal-in-Space Interface Control Document, Issue 2.0*. European Union
12. Paziewski J., & Wielgosz P. (2014). Accounting for Galileo–GPS intersystem biases in precise satellite positioning. *Journal of Geodesy, 89*(1), 81–93.
13. Reckeweg, F. (2020). Integer ambiguity resolution for multi-GNSS and multi-signal raw phase observations. ISBN 978-3-935631-47-1.
14. Roskosmos. (2021). GLONASS constellation status. glonass-iac.ru/en. Accessed September 2021.

Part II
Key Elements of GNSS Geodesy

This part, comprising of five chapters, introduces a selection of key elements of the theory and practice of GNSS geodesy to help understand the concepts of achieving high precision from GNSS data. Discussions of existing models and strategies are provided with the assumption that the reader is primarily interested in understanding the concepts applicable in achieving high precision from GNSS positioning. Pertinent references are provided throughout as necessary for further reading. The chapters are designed as follows:

Chapter 3 contains a review of reference systems (such as time scales, coordinate systems, reference surfaces, GNSS-derived heights, earth orientation, and reference frames) as used in the practice of GNSS geodesy.

Chapter 4 deals with the concept of geodetic parameter estimation from GNSS. There are several geodetic quantities (parameters) that can be derived (estimated) using precise measurements and geodetic science, for which GNSS technologies are one of the widespread, globally accessible tools.

Chapter 5 is devoted to GNSS measurement and observation models. It reviews the defining observation equations for GNSS signals, signal propagation errors, and correction methods for achieving precise positioning.

Chapter 6 covers the geodetic GNSS CORS networks that are the ground infrastructure for recording reference data for high-precision processing and realizations of reference frames from reference system definitions.

Chapter 7 introduces data processing strategies, including preprocessing, ambiguity fixing for precise positioning, and solution reprocessing methods as commonly used in both epoch solutions and time series data.

Chapter 3
Reference Systems in GNSS Geodesy

3.1 Time Scales and Their Relationships

Time scales are *time reference systems* used for timing and timekeeping [2]. The dictionary definition of timing is *"to measure the time taken by. . .or to record time of. . ."* and timekeeping is *"the process or activity of timing an event or series of events."*

GNSS satellites have atomic clocks onboard that are used to construct and disseminate time scales such as GPS Time, GLONASS Time, Galileo System Time, and BeiDou Time. These time scales and how they relate to each other and to other time scales in GNSS geodesy, such as TAI, UTC, and UT1, are the primary focus of this section.

The construction of time scales is based on either astronomical observations or physical timing instruments (clocks). The concept of a clock system, including the rotating Earth and movement of celestial bodies, involves (1) a periodic movement that can be observed, (2) the continuous counting of the periods, and (3) the display of the registered count. *Free spinning rotors* and *harmonic oscillators* are among the periodic physical movements used to define practical time scales. The Earth is an example of a free spinning rotor and Earth's rotation as observed by astronomical or geodetic means defines the UT1 time scale. Pendulum, quartz crystal, and atomic resonators are examples of harmonic oscillators which lead to a variety of clocks and time scales. Atomic clocks are the most accurate and lead to the concept of atomic time scales such as TAI and various GNSS time scales.

Starting from a clock system (using the abovementioned periodic movements), a time scale is obtained by two main actions, namely (1) the *period*, or its inverse (the frequency), of the basic oscillation must be measured, adjusted, or defined and (2) the *origin* of the time scale must be specified. *The first action, which consists of establishing a unit of time interval, is not sufficient to define the time scale. Only the choice of origin from which we start counting the periods completes the task* [18]. In practice, both actions require conventions to be defined and/or agreed

C. A. Ogaja, *Introduction to GNSS Geodesy*,
https://doi.org/10.1007/978-3-030-91821-7_3

upon, resulting in what is also commonly referred to as time (frequency) standards. Uniformity and stability of a time scale are the most important requirements.

Case in Point is the Earth-orbiting GPS satellites, each with atomic clocks on board, synchronized to a single time scale, enabling precise timing for accurate positioning. A precise timing device in a GPS receiver is used by its computer to calculate the time of flight of the signal from each of the observable satellites (see details in Chaps. 4 and 5). This concept is characteristic of all GNSS systems. Thus, without knowing the relationships between GNSS time scales, it would be impossible to effectively combine measurements from different GNSS constellations to obtain a position.

Time Scales Based on Earth Rotation: Historically, timing was tied to sunrise, sunset, and the seasons, and astronomical observations provided both parts of the clock: the frequency standard (one cycle per day, obtained from the Earth's spin[1]) and the counter (calendar). The following time scales are based on Earth rotation.

Apparent Solar Time is based on solar day, which is the interval between two transits of the Sun across a local meridian. A solar day is approximately 24 hours of mean time but is not uniform throughout the year due to irregularities in Earth's spin rate, non-ellipticity of Earth's orbit around the Sun, and Earth's polar wobble, among other perturbations.

Mean Solar Time is obtained by averaging the actual (or apparent) solar time over one year. This cancels the variations in the apparent solar time due to the aforementioned irregularities and perturbations.

Sidereal Time is time by the stars. A sidereal day is the interval between two transits of the same star across the local meridian and is approximately 23 hours 56 minutes 4 seconds long. Thus, a sidereal day is 4 minutes shorter than a solar day (from the mean solar time).

Universal Time (UT) is the mean solar time at the Greenwich meridian (0° longitude). It replaced GMT which used to be an international time standard until 1928. **UT1**, a refined version of UT, is derived in two steps. First, the sidereal time at a given location on the Earth is converted to mean solar time and referred to the Greenwich meridian to obtain UT0. **UT1** is then computed by correcting UT0 for the effect of polar motion on the longitude of the observing site. Applying corrections for seasonal variations of the Earth's spin rate results in UT2; however, UT2 is not significantly more uniform than UT1 over long time intervals.

Some Practical Considerations: Although GNSS satellites have atomic clocks on board, the concept of time and its passage as determined from Earth rotation (e.g., day, hour, minute) plays a key role in the design of GNSS orbits (e.g., in defining orbital periods). For example, relative to the Earth,

1. Each **GPS** satellite **orbits twice in a sidereal day**, the length of time it takes a star (not the sun) to return to the same position in the sky, as seen by a stationary

[1] We all have 86,400 seconds a day.

observer on the Earth. Sidereal day is four minutes shorter than a solar day, and therefore each GPS satellite orbits the Earth once every 11 hours and 58 minutes.
2. Each **GLONASS** satellite completes the orbit in approximately 11 hours 15 minutes. This means that for a stationary observer the same satellite is visible at the same point in the sky every **eight sidereal days**.
3. Each **Galileo** satellite completes its orbit in 14 hours 5 minutes. After **ten sidereal days**, the orbit of any Galileo satellite repeats itself.

Navigators, astronomers, and the like need to know the Earth's angular position and the Earth time (**UT1**), which in reality is the mean time at Greenwich. However, precise time and frequency users such as geodesists, scientists, and engineers are satisfied with the excellent performance of atomic clocks. It is known and accepted that atomic clocks are better for defining accurate and reliable time intervals through harmonic resonance. Because of this and the need to have and distribute the most uniform and most accurate time possible, it was necessary to develop agreed upon relationships between **UT1** and atomic time scales such as **UTC** (discussed later).

Atomic Time Scales: Atomic time is a time scale obtained by continuous counting of SI seconds.[2]

International Atomic Time (TAI) is the atomic time scale kept by the International Bureau of Weights and Measures (BIPM). Its unit interval is exactly one SI second at sea level, and its origin is such that UT1-TAI was 0h 0m 0s on January 1, 1958 [3]. TAI time scale is based on the combined input of hundreds of highly precise atomic clocks around the world, each corrected for environmental and relativistic effects.

Universal Time Coordinated (UTC), despite its name, is also atomic time scale. It is simply the TAI *plus* a time varying offset. Therefore, both the TAI and UTC are from the same combined input (e.g., as weighted average of datasets) from the hundreds of highly precise atomic clocks from around the world. Mathematically, UTC = TAI−O(t), where O is the offset in seconds at epoch t (e.g., 1 Jan 1972, 1 Jan 2017, and so forth), and the offset O is negative because TAI is designed to be ahead of UTC. *But why use UTC and why the offset?*.[3] While both time scales are defined from the same datasets, TAI is kept as determined from those datasets, but UTC (introduced in 1972 as the world's official time, because of the need to use an atomic time scale that also takes into account the Earth's slowing rotation, which determines the length of a day) is adjusted to be within 0.9 seconds of **UT1**. This is done by introducing a **leap second**[4] to the UTC time before the 0.9 second threshold

[2] SI second is the time it takes a Cesium-133 atom at the ground state to oscillate exactly 9,192,631,770 times [4].

[3] The continuing requirement for a time scale approximating UT is due to its wide application in surveying and navigation. A compromise solution had, therefore, to be found which retains the advantage of uniform time scale generation by atomic clocks and still follows the variations of the Earth's rotation [18].

[4] A one-second adjustment that is occasionally applied to UTC in order to keep its time of day close to the mean solar time as realized by UT1.

is reached. Thus, the relationships between UTC, TAI, and UT1 time scales are defined as follows (e.g., [2, 4]):

$$\begin{aligned} TAI &= UTC + 1^s \times n \\ UTC &= UT1 + dUT1, \qquad |dUT1| < 0.9^s \end{aligned} \tag{3.1}$$

where n is the number of leap seconds introduced for a given epoch (e.g., for 1 Jan 1972, $n = 10$, 1 Jan 1999, $n = 32$, 1 Jan 2006, $n = 33$, 1 Jan 2009, $n = 34$, 1 Jul 2012, $n = 35$, and 1 Jan 2017, $n = 37$). Therefore, as of 31 December 2016, when another **leap second** was added, TAI was ahead of UTC by 37s. The 37s results from the initial difference of 10 seconds at the start of 1972, plus 27 leap seconds in UTC from that time to 31 December 2016. If additional leap seconds are added to UTC in future, this relationship will also change, i.e., in terms of the number of seconds TAI is ahead of UTC.

Estimates of UTC are computed and provided by different entities such as UTC(USNO) from the United States Naval Observatory (USNO), UTC(NIST) from the National Institute of Standards and Technology (NIST), and many others. In general, UTC(k) is a realization of UTC by a given laboratory k, see [2].[5] The UTC(k) data from around the world are communicated to the BIPM for the latter's calculation of TAI and UTC.

GPS Time (GPST) [13] starts at 0h UTC (midnight) of 6 January 1980 and has no leap second adjustments, with **TAI−GPST** set at **19 seconds**. Hence **GPST−UTC** = $n -$ **19 s**. GPS is not only a global source of time but also provides the means of transferring time from one location to another. And in fact, it can be said that three kinds of time are available from GPS [2]:

> *GPS time*, *UTC(USNO)* as estimated and produced by the United States Naval Observatory, and *the times from each GPS satellite's atomic clock*. The Master Control Station (MCS) in Colorado Springs, Colorado, USA, gathers the GPS satellites' data from the monitor stations around the globe. A Kalman filter software program estimates the time error, frequency error, frequency drift and Keplerian orbit parameters for each of the satellites and its operating clock. This information is uploaded to each satellite so that it can be broadcasted in real time. This process provides GPS time consistency across the constellation to within a small number of nanoseconds.

GPST as defined in GPS navigation messages (Sect. 6.3.1) uses 10 bits to convey GPS week numbers (starting with week 0 on 6 Jan 1980). The 10-bit representation only covers a range of zero to 1023, a duration of 1024 GPS weeks (19.7 years). However, GPS time is a continuous time scale, and thus *GPS week counter* would *roll-over* from 1023 to zero at the end of the 19.7 year cycle (e.g., few seconds after

---

[5] The work to generate UTC is performed at the BIPM in France. The staff doing the work is composed of several international timing experts who frequently interact with, and obtain timing data from, the rest of the world's time and frequency community. The leap second steps are determined by the IERS, which operates out of the Paris Observatory and which collects the Earth's rotation data from numerous observatories and radio telescopes around the globe.

Table 3.1 GPS week rollover cycles

GPS week cycle	Start of week 0	End of week 1023
1	January 6, 1980 (44244)	August 21, 1999 (51411)
2	August 22, 1999 (51412)	April 6, 2019 (58579)
3	April 7, 2019 (58580)	November 20, 2038 (65747)

Note: The numbers in parentheses are the corresponding modified Julian date (MJD) (Days elapsed since midnight, November 17, 1858, where MJD = Julian date (JD) − 2,400,000.5, JD = int[365.25 $\times y$] + int[30.6001 $\times$ (m+1)] + DD + UT(h)/24 + 1,720,981.5. (y = YY−1, m = MM+12 for MM ≤ 2, else y = YY, m = MM))

midnight, UTC, on 7 April 2019). This has been known to cause some challenges[6] although it had been suggested (as of 26 September 2017) that modernized GPS navigation message would have a 13-bit week number in order to resolve them [15]. Table 3.1 shows the first three GPS week rollover cycles since the start of GPST.

GLONASS Time (GLNT) [12] is a time scale generated by GLONASS Central Synchronizer. It is 3 h ahead of UTC time and introduces leap second adjustments just like the UTC time scale. Hence it is different from the other GNSS time scales in terms of the leap second adjustment, as GPS, Galileo, and BeiDou time scales do not introduce leap seconds. The offset between GLNT and UTC(SU)[7] should not exceed 3 h plus 1 ms (i.e., **GLNT=UTC+3h-**τ, where $|\tau| < 1$ ms), and the relationship between GLNT and GPST is given by **GLNT=GPST+(19s−**n**)+3h−**τ, where n is previously explained in Eq. 3.1. Inter-GNSS time scale differences and offsets are applied during positioning. For example, *in a combined GPS/GLONASS mode, the receiver must track a minimum of five satellites (e.g., four GPS satellites representing four unknowns of 3D position and time and at least one GLONASS satellite to determine the GPS/GLONASS time offset).*

Galileo System Time (GST) [8] starts at 0h UTC, 22 Aug 1999, has no leap second adjustments, and **GST−UTC** = $n-$ **19s**. Therefore, apart from the origin, its definition is similar to that of GPST. The origin of GST coincided with the end of first (and start of second) GPS week rollover cycle (Table 3.1). Defined within the Galileo navigation message (Sect. 6.3.1) as a 32-bit number, GST is composed of two parameters, namely, Galileo week number (12 bits) and time of week[8] (20 bits). The 12-bit week number covers 4096 weeks (about 78 years) from the origin, after which the counter resets to zero to cover additional period modulo 4096. The

[6] Potential *GPS week rollover* issues would include (1) UTC timing displayed and/or time tags of receiver data containing PNT information could jump by 19.7 years and (2) associated time tags of GPS solutions could be incorrect, causing problems such as GPS receiver engine failure or loss of data.

[7] The national estimate of UTC generated by the timing entity in Russia.

[8] Galileo/GPS time of week is the number of seconds that have occurred since the transition from the previous week. It covers an entire week from 0 to 604,799 seconds and is reset to zero at the end of each week.

Galileo week zero corresponds to GPS week 1024 (first cycle), which after week roll-over was reported as week zero.

BeiDou Time (BDT) [20] starts at 0h UTC, 1 January 2006, has no leap second adjustments, and is 33 seconds behind TAI (i.e., **TAI−BDT=33s**). Hence **BDT−UTC** $= n -$ **33s**. The BeiDou navigation message [1] conveys BDT in BeiDou week number (13 bits) and seconds of week[9] (20 bits). The 13-bit week number is the integral week count of BDT with the range of zero through 8191 (i.e., 8192 weeks, which is about 157.5 years). The BDT week number started from zero at 0h (midnight) of UTC on 1 January 2006, and based on the duration of its cycle (157.5 years), there would be no need for a BDT week roll-over, in a very long time.

3.2 Geodetic Datums and Coordinate Systems

Geodetic positions are given in terms of coordinates, a set of numbers that locate the point of interest relative to a reference point (origin). Four coordinates are needed, three to define position relative to a set of axes attached to the Earth's figure in some prescribed manner and the fourth to mark the time at which the coordinates were determined. The time or epoch at which the reference axes were determined is also important.

GNSS satellite orbits (theory in Appendix A) are centered at the Earth's center of mass, the geocenter. Therefore, naturally the geocenter becomes an important reference point (of beginning), the origin, for defining a global reference (datum) and/or coordinate system for the practice of geodesy using GNSS satellites. Here, we describe the coordinate types and reference systems which occur in GNSS geodesy and show how coordinates expressed in one system may be transformed to coordinates in another.

3.2.1 Geocentric Cartesian Coordinates

The fact that GNSS satellites are geocentric around geocenter (the center of Earth's gravity) necessitated the development of a coordinate system appropriate for global use. Such a coordinate system would be based on a geodetic datum (a mathematical surface, an ellipsoid, used to approximate the Earth's shape and size). Since the satellite systems are global in coverage, the center of the ellipsoid and the center

[9] BeiDou seconds of week is the number of seconds that have occurred since the last Sunday, 00:00:00 BDT. It covers an entire week from 0 to 604,799 seconds and is reset to zero at the end of each week.

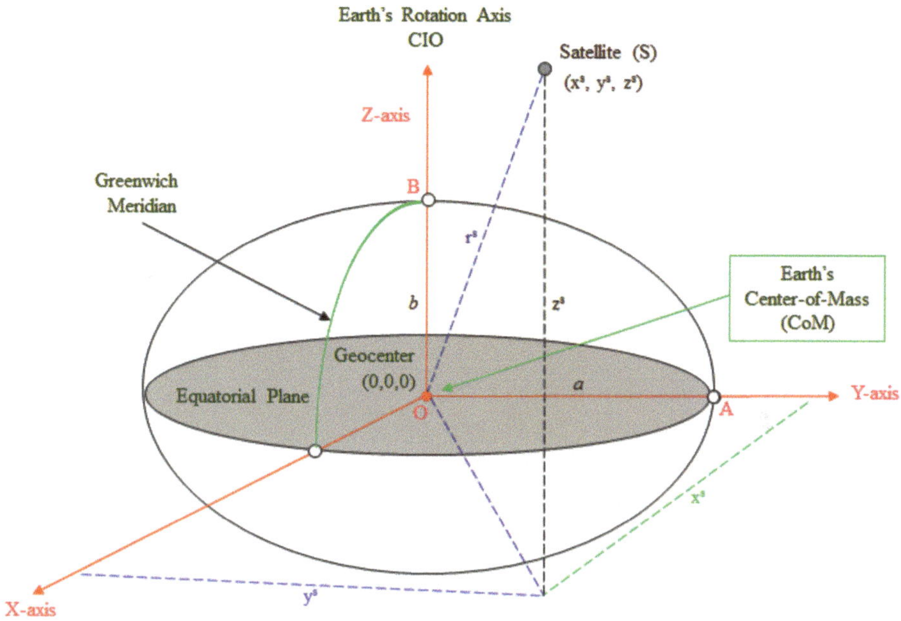

Fig. 3.1 Geocentric Cartesian coordinate system

of the geoid (Earth's gravitational mass) would coincide, and hence the ellipsoid is geocentric around the center of Earth's gravity (or geoid), just as the orbits of the satellites are.

Geocentric Cartesian coordinates of a point on the Earth or of an object in space are defined in an Earth-centered (geocentric), Earth-fixed, orthogonal, three-dimensional axis system. For example, in Fig. 3.1, the geocentric Cartesian coordinates (x^S, y^S, z^S) of satellite S at a time instant in space are defined as offsets from the origin (0, 0, 0) along each of the three orthogonal axes. This system is sometimes referred to as the ECEF (Earth-centered Earth-fixed) system. The origin of the coordinate system is at the Earth's center of mass (CoM) that corresponds with the center of the ellipsoid (intersection of equatorial plane and the Earth's spin axis, CIO). The X-axis lies in the equatorial plane with its positive end intersecting the Greenwich meridian. The Y-axis lies in the equatorial plane with its positive end intersecting the ellipsoid at 90° East longitude. The Z-axis is coincident with the Earth's rotation axis and is considered positive toward the North Pole, also referred to as the Conventional International Origin (CIO). The distance, r^S, from the geocenter to the satellite position is also illustrated.

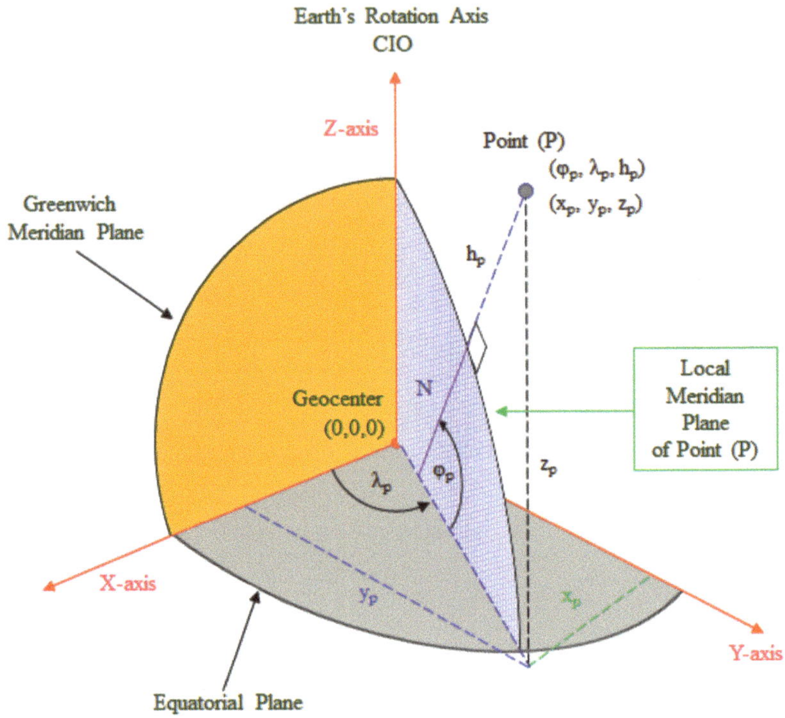

Fig. 3.2 Geodetic (ellipsoidal) coordinate system

3.2.2 Latitude, Longitude, and Ellipsoidal Height

A point on or above the Earth's surface can be located by its Cartesian coordinates (Sect. 3.2.1) or ellipsoidal coordinates (aka geodetic coordinates). Figure 3.2 illustrates both the geocentric Cartesian coordinates and the geodetic coordinates for a point P located above the ellipsoid surface. The latter are expressed using geodetic latitude (φ_p), geodetic longitude (λ_p), and height above or below the ellipsoid surface (h_p). The longitude is expressed as positive to the east of the Greenwich meridian. Latitude is the angle measured from the equatorial plane to the normal N, a line passing through the point P and normal (perpendicular) to the ellipsoid surface at the point of intersection with that surface. Latitude is expressed as positive to the north. Ellipsoidal height is expressed as positive above the ellipsoid surface.

3.2.3 Local Geodetic Horizon Coordinates

Local geodetic horizon coordinates (also known as Local Topocentric coordinates) are useful for hybrid geodetic positioning integrating GNSS-derived positions with terrestrial measurements taken with optical instruments (e.g., total station). The local geodetic horizon coordinate system has its origin at any point specified. For example, in Fig. 3.3, an Earth-fixed, right-handed orthogonal three-dimensional e, n, u coordinate system is shown for point P where two axes (e and n) lie on a local horizon plane which is tangent at the point, and the third axis is normal to the horizon plane. The north axis (n) lies in the local meridian plane of point P and is positive toward the North Pole (CIO). The up axis (u) lies along a normal to the ellipsoid at the point and is positive above the horizon surface at the origin (in this case point P). The east axis (e) completes the right-handed system by being perpendicular to the local meridian plane and is positive to the east.

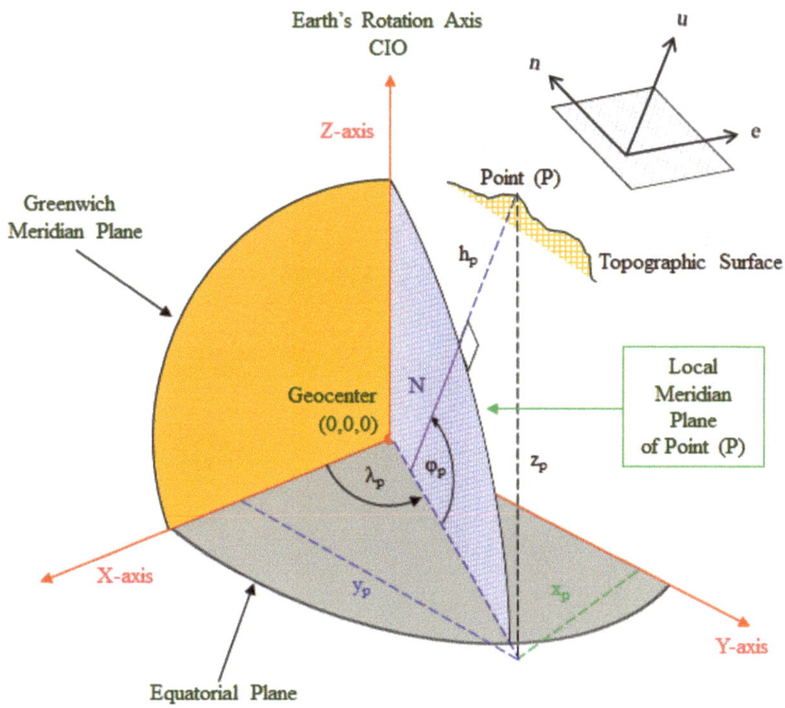

Fig. 3.3 Local geodetic horizon coordinate system

3.2.4 Coordinate System Conversions

3.2.4.1 Cartesian and Ellipsoidal Coordinate Conversions

Geocentric Cartesian coordinates (x, y, z) can be obtained from geodetic (ellipsoidal) coordinates (φ, λ, h) using the following expressions:

$$
\begin{aligned}
x &= (N + h)cos\varphi cos\lambda \\
y &= (N + h)cos\varphi sin\lambda \\
z &= ((1 - e^2)N + h)sin\varphi
\end{aligned}
\tag{3.2}
$$

where N is the radius of curvature in the prime vertical (Fig. 3.2)

$$N = \frac{a}{\sqrt{1 - e^2 sin^2\varphi}} \tag{3.3}$$

and the eccentricity e is related to the semi-major axis a, the semi-minor axis b, and the flattening factor f, by

$$e^2 = 2f - f^2 = \frac{a^2 - b^2}{a^2} \tag{3.4}$$

The ellipsoidal (geodetic) coordinates (φ, λ, h) can be derived from geocentric Cartesian coordinates (x, y, z) as follows:

The longitude λ is given by

$$\lambda = arctan\left(\frac{y}{x}\right) \tag{3.5}$$

Height h and latitude φ are given by

$$h = \frac{\sqrt{x^2 + y^2}}{cos\varphi} - N \tag{3.6}$$

and

$$\varphi = arctan\left[\frac{z}{\sqrt{x^2 + y^2}}\left[1 - e^2\left(\frac{N}{N + h}\right)\right]^{-1}\right] \tag{3.7}$$

Iteration is required for φ and h because they are inter-dependent in 3.6 and 3.7. This is done by first assuming $h = 0$ and solving for φ. Then solving for h and then again for φ. This continues until the change in φ between successive iterations is smaller than the required precision. Thus,

1. The initial value of latitude is given by

$$\varphi_{(0)} = arctan\left[\frac{z/(1-e^2)}{\sqrt{x^2+y^2}}\right] \tag{3.8}$$

2. Subsequent improved values are computed, whereby

$$\begin{aligned} N_{(i)} &= \frac{a}{\sqrt{1-e^2 sin^2\varphi_{(i-1)}}} \\ h_{(i)} &= \frac{\sqrt{x^2+y^2}}{cos\varphi_{(i-1)}} - N_{(i)} \end{aligned} \tag{3.9}$$

are used in 3.7 to compute $\varphi_{(i)}$, and the iterations are repeated until the difference between $\varphi_{(i)}$ and $\varphi_{(i-1)}$ is either insignificant or smaller than the required precision.

3.2.4.2 ECEF and ENU Coordinate Conversions

Given the ECEF Cartesian coordinates of a point P (Fig. 3.3) and its associated latitude and longitude, a displacement vector (Δx, Δy, Δz,) from that point can be converted from the ECEF Cartesian system to the local system (Δe, Δn, Δu,) and vice versa.

1. Converting the displacement vector from ENU to ECEF coordinates is accomplished using

$$\begin{bmatrix} \Delta x \\ \Delta y \\ \Delta z \end{bmatrix} = \begin{bmatrix} -sin\lambda & -cos\lambda sin\varphi & cos\lambda cos\varphi \\ cos\lambda & -sin\lambda sin\varphi & sin\lambda cos\varphi \\ 0 & cos\varphi & sin\varphi \end{bmatrix} \begin{bmatrix} \Delta e \\ \Delta n \\ \Delta u \end{bmatrix} \tag{3.10}$$

2. Converting the displacement vector from ECEF to ENU coordinates is accomplished using

$$\begin{bmatrix} \Delta e \\ \Delta n \\ \Delta u \end{bmatrix} = \begin{bmatrix} -cos\lambda & sin\lambda & 0 \\ -cos\lambda sin\varphi & -sin\lambda sin\varphi & cos\varphi \\ cos\lambda cos\varphi & sin\lambda cos\varphi & sin\varphi \end{bmatrix} \begin{bmatrix} \Delta x \\ \Delta y \\ \Delta z \end{bmatrix} \tag{3.11}$$

3.2.4.3 Satellite Elevation and Azimuth Computation

Satellite elevation (E) and azimuth (A) at a given point P are defined by the local horizon system as shown in Fig. 3.4. The receiver coordinates at that point and the

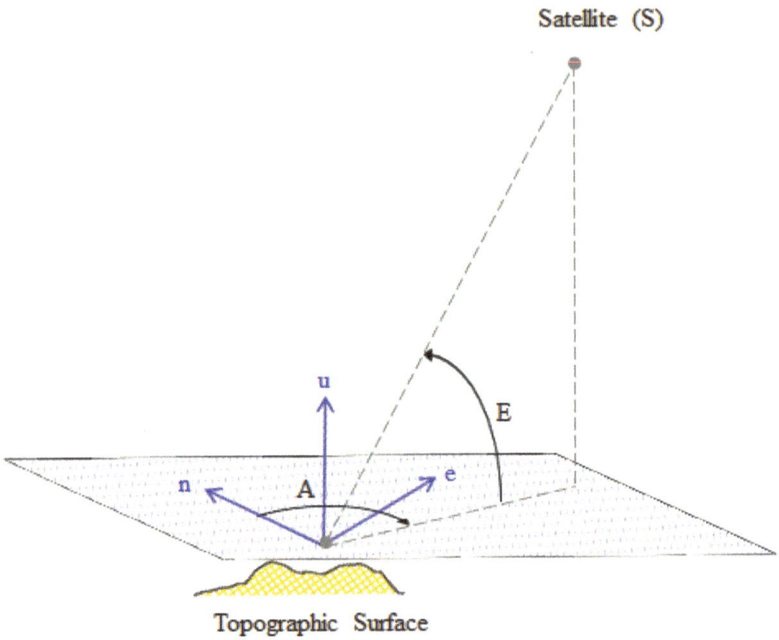

Fig. 3.4 Local coordinate frame showing satellite azimuth (A) and elevation (E)

corresponding satellite coordinates are applied to compute the satellite elevation and azimuth.

1. Given the line-of-sight unit vector (Fig. 3.5),

$$\hat{\rho} = \frac{\boldsymbol{r}^{sat} - \boldsymbol{r}_{rcv}}{||\boldsymbol{r}^{sat} - \boldsymbol{r}_{rcv}||} = \frac{\Delta \boldsymbol{r}}{\rho} \tag{3.12}$$

where $\boldsymbol{r}^{sat}$ and $\boldsymbol{r}_{rcv}$ are the geocentric Cartesian position of the satellite and receiver, respectively, and

$$\Delta \boldsymbol{r} = \begin{bmatrix} \Delta x \\ \Delta y \\ \Delta z \end{bmatrix} \tag{3.13}$$

2. Rotate 3.12 from XYZ to ENU using the following:

$$\begin{bmatrix} \hat{\rho} \cdot \hat{e} \\ \hat{\rho} \cdot \hat{n} \\ \hat{\rho} \cdot \hat{u} \end{bmatrix} = \begin{bmatrix} -cos\lambda & sin\lambda & 0 \\ -cos\lambda sin\varphi & -sin\lambda sin\varphi & cos\varphi \\ cos\lambda cos\varphi & sin\lambda cos\varphi & sin\varphi \end{bmatrix} \begin{bmatrix} \Delta x/\rho \\ \Delta y/\rho \\ \Delta z/\rho \end{bmatrix} \tag{3.14}$$

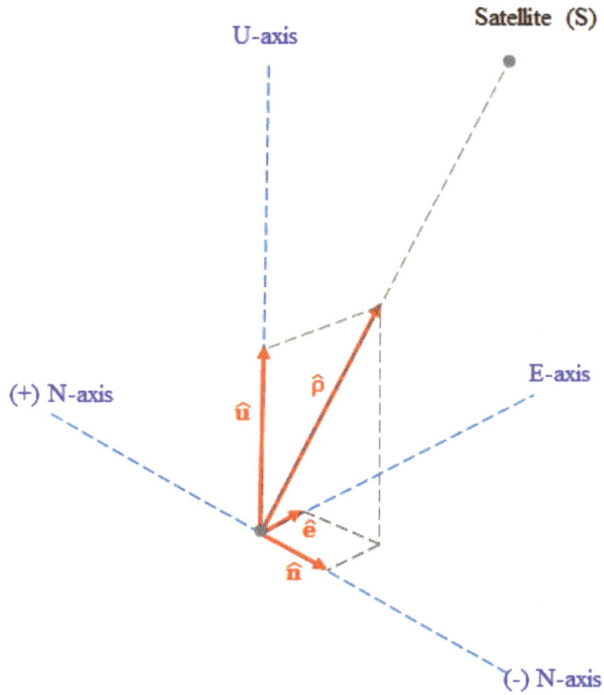

Fig. 3.5 Unit vector of receiver-satellite range

3. Compute the elevation and azimuth in the local coordinate system using

$$E = arcsin(\hat{\rho} \cdot \hat{u}) \tag{3.15}$$

$$A = arctan\left(\frac{\hat{\rho} \cdot \hat{e}}{\hat{\rho} \cdot \hat{n}}\right) \tag{3.16}$$

3.2.5 Datum Transformations and Map Projections

When using GNSS, coordinates are obtained in a global geocentric datum.[10] A geocentric datum is defined when physical points located on the Earth's surface

[10] A datum is defined when points are positioned with reference to a particular ellipsoid whose shape, size, position, and orientation relative to the Earth's surface are known. The position and orientation of an ellipsoid may be defined using physical point(s) located on the Earth's surface, through assigned latitude and longitude.

have their geodetic (ellipsoidal) coordinates referred to a particular ellipsoid whose shape and size are known and has the same origin as the geocentric Cartesian coordinate system (Fig. 3.1). However, different geocentric datums may share the same geocentric Cartesian coordinate system but apply different ellipsoid parameters or have same ellipsoidal parameters but different realizations over time. There may also be instances where a non-geocentric[11] datum is the best choice for a particular region, country, or application. Due to these reasons, datum transformations are required when working with GNSS results on different datums or reference epochs. Furthermore, geodetic results are usually preferred as ellipsoidal coordinates, local coordinates, or grid plane coordinates. Producing results in grid plane coordinates requires a map projection of the datum coordinates.

3.2.5.1 Datum Transformations

Datum transformations vary from a simple three-parameter transformation to a complex seven-parameter transformation with 14-term adjustments. The seven-parameter similarity transformation, also called the Helmert transformation, is used for reference systems provided by satellite and space geodetic techniques. In its basic form, it applies one scale factor, three translations, and three rotations. However, when dealing with datums from multi-epoch or multi-year datasets, it is necessary to augment the classical seven parameters with their time derivatives [5, 17]. This extends the seven-parameter Helmert transformation to 14-term adjustment formulations to better incorporate time-dependent processes such as plate tectonics and other geophysical phenomena. The pertinent concepts of the Helmert transformation can be briefly described as follows:

1. For two datums A and B, the general form is

$$\begin{bmatrix}\varphi\\ \lambda\\ h\end{bmatrix}_{\mathrm{A}} \Leftrightarrow \begin{bmatrix}x\\ y\\ z\end{bmatrix}_{\mathrm{A}} \leftarrow \textbf{Transformation} \rightarrow \begin{bmatrix}x\\ y\\ z\end{bmatrix}_{\mathrm{B}} \Leftrightarrow \begin{bmatrix}\varphi\\ \lambda\\ h\end{bmatrix}_{\mathrm{B}} \tag{3.17}$$

2. Consider three translation components, scale (dilation) factor, and three rotation angles, denoted, respectively, as Tx, Ty, Tz, D, Rx, Ry, Rz, and their time derivatives $\dot{T}\mathrm{x}$, $\dot{T}\mathrm{y}$, $\dot{T}\mathrm{z}$, $\dot{D}$, $\dot{R}\mathrm{x}$, $\dot{R}\mathrm{y}$, $\dot{R}\mathrm{z}$. The transformation of coordinate vector $\boldsymbol{X}_A$, expressed in datum A, into a coordinate vector $\boldsymbol{X}_B$, expressed in datum B, is formulated as

$$\boldsymbol{X}_B = \boldsymbol{X}_A + \boldsymbol{T} + D\boldsymbol{X}_A + \boldsymbol{R}\boldsymbol{X}_A \tag{3.18}$$

[11] A datum defined using an ellipsoid whose center and the origin of its associated Cartesian coordinate system coincide but are non-geocentric.

where

$$\boldsymbol{T} = \begin{bmatrix} Tx \\ Ty \\ Tz \end{bmatrix} \text{ and } \boldsymbol{R} = \begin{bmatrix} 0 & -Rz & Ry \\ Rz & 0 & -Rx \\ -Ry & Rx & 0 \end{bmatrix} \tag{3.19}$$

The assumed units for the parameters are meter for positions and translations, 10^{-6} for scale and arc seconds for rotations. Since generally $\boldsymbol{X}_A$, $\boldsymbol{X}_B$, $\boldsymbol{T}$, D, $\boldsymbol{R}$ are a function of time, differentiating Eq. 3.18 with respect to time gives

$$\dot{\boldsymbol{X}}_B = \dot{\boldsymbol{X}}_A + \dot{\boldsymbol{T}} + \dot{D}\boldsymbol{X}_A + D\dot{\boldsymbol{X}}_A + \dot{\boldsymbol{R}}\boldsymbol{X}_A + \boldsymbol{R}\dot{\boldsymbol{X}}_A \tag{3.20}$$

For global datums established by GNSS and space geodetic techniques, and $\boldsymbol{T}$ is less than 100 meters, D and $\boldsymbol{R}$ are less than 10^{-5} [5]. $D\dot{\boldsymbol{X}}_A$ and $\boldsymbol{R}\dot{\boldsymbol{X}}_A$ in Eq. 3.20 are negligible (0.0 mm per 100 years), thus

$$\dot{\boldsymbol{X}}_B = \dot{\boldsymbol{X}}_A + \dot{\boldsymbol{T}} + \dot{D}\boldsymbol{X}_A + \dot{\boldsymbol{R}}\boldsymbol{X}_A \tag{3.21}$$

The corresponding units for the time derivatives are meter per year for position velocities, meter per year for translation rates, 10^{-6} per year for scale rate, and arc seconds per year for rotation rates.

3. Equation 3.18 is a simpler linearized version of a more complex derivation (e.g., as found in [5, 11, 17]). If the seven transformation parameters are not known beforehand, their estimation can be achieved through a least squares process [9, 10], in which a minimum of seven equations must be written for a unique solution. This requires a minimum of three control points with x, y, z coordinates in both datums although in reality, tens or hundreds of control points are usually used. For N control points ($i = 1, \ldots, N$), the standard nonlinear relation between the two datums A and B would be

$$\boldsymbol{X}_B = \boldsymbol{T} + s\boldsymbol{R}'\boldsymbol{X}_A \tag{3.22}$$

where $\boldsymbol{R}'$ is a nonlinear 3×3 matrix whose individual elements are sine and cosine functions of the three rotation angles Rx, Ry, and Rz (see, e.g., [9, page 358]). Global and GNSS datums such as ITRF and WGS 84 (discussed later in Sects. 3.5.1 and 3.5.2) use the same definitions for the axes and origin but differ slightly due to their actual realizations, e.g., through different networks. The rotations and translations between such datums are usually very small and the scale factor should be nearly 1, and given that the sine of a very small angle is equal to the angle in radians and its cosine is nearly 1, the matrix $\boldsymbol{R}'$ is simply defined as

$$\boldsymbol{R}' = \boldsymbol{I} + \begin{bmatrix} 0 & -Rz & Ry \\ Rz & 0 & -Rx \\ -Ry & Rx & 0 \end{bmatrix} = \boldsymbol{I} + \Delta\boldsymbol{R} \tag{3.23}$$

Similarly, the translation parameters and the scale factor can be modified as follows:

$$\boldsymbol{T} = \boldsymbol{T}_0 + \Delta\boldsymbol{T} \tag{3.24}$$

$$s = 1 + D \tag{3.25}$$

In Eq. 3.24, the approximate shift vector $\boldsymbol{T}_0$ can be computed from a single common point, as the difference between the two datum coordinates, i.e., $\boldsymbol{T}_0 = \boldsymbol{X}_B - \boldsymbol{X}_A$, or an average of all the common points can be used to obtain initial approximations such that

$$\boldsymbol{v} = \boldsymbol{X}_B - \boldsymbol{X}_A - \boldsymbol{T}_0 \tag{3.26}$$

For a single point i, the linearized model for Eq. 3.22 becomes

$$\begin{bmatrix} x_i \\ y_i \\ z_i \end{bmatrix}_B - \begin{bmatrix} x_i \\ y_i \\ z_i \end{bmatrix}_A - \boldsymbol{T}_0 = \boldsymbol{J}_i \Delta\boldsymbol{u} \tag{3.27}$$

$$\Delta\boldsymbol{u} = (\Delta Tx, \Delta Ty, \Delta Tz, D, Rx, Ry, Rz)^T \tag{3.28}$$

$$\boldsymbol{J}_i = \begin{bmatrix} 1\,0\,0\; x_i & 0 & z_i & -y_i \\ 0\,1\,0\; y_i & -z_i & 0 & x_i \\ 0\,0\,1\; z_i & y_i & -x_i & 0 \end{bmatrix}_A \tag{3.29}$$

For N control points ($i = 1, \ldots, N$), the least squares adjustment gives the best estimate for $\Delta\boldsymbol{u}$:

$$\Delta\boldsymbol{u} = (\boldsymbol{J}^T\boldsymbol{J})^{-1}\boldsymbol{J}^T\boldsymbol{v} \tag{3.30}$$

4. Similarly, if the time derivatives (in Eq. 3.21) of the seven parameters are not known, they can be estimated from a least squares solution using

$$\Delta\dot{\boldsymbol{u}} = (\boldsymbol{J}^T\boldsymbol{J})^{-1}\boldsymbol{J}^T\dot{\boldsymbol{v}} \tag{3.31}$$

where

$$\Delta\dot{\boldsymbol{u}} = (\Delta\dot{T}x, \Delta\dot{T}y, \Delta\dot{T}z, \dot{D}, \dot{R}x, \dot{R}y, \dot{R}z)^T \tag{3.32}$$

$$\dot{\boldsymbol{v}} = \dot{\boldsymbol{X}}_B - \dot{\boldsymbol{X}}_A - \dot{\boldsymbol{T}}_0 \tag{3.33}$$

5. Consider further a special case in which coordinates on a fixed datum are moving at a certain rate in space with respect to that same datum that remains fixed [17]. The coordinates and their linear velocities are known at some arbitrary epoch t_0, and new coordinates are to be computed at epoch t. For this, Eq. 3.22 is modified to take the form:

$$\boldsymbol{X}_{yy}(t) = \boldsymbol{T}(t) + s\boldsymbol{R}'\{\boldsymbol{X}_{00}(t_o) + (t - t_0)\boldsymbol{V}_{00}\} \tag{3.34}$$

where $\boldsymbol{X}_{yy}$ and $\boldsymbol{X}_{00}$ are datum coordinates at the two epochs (for example, $ITRF_{yy}$ and $ITRF_{00}$) and $\boldsymbol{V}_{00}$ is the vector of linear velocities attached to the datum coordinates $\boldsymbol{X}_{00}$ at epoch t_0.

The 14-term transformation parameter formulations are based on Eqs. 3.18, 3.21, 3.30, and 3.31. These concepts are discussed further in Sect. 7.4.3 of Chap. 7 where time series data combinations and terrestrial reference frame realizations are addressed.

3.2.5.2 Map Projections

Geodetic coordinates (latitude, longitude, and height on an ellipsoidal model of the Earth) are three-dimensional (3D), but maps and GIS display mapped information in a two-dimensional (2D) grid coordinates system. Some of the practical necessities for such a system include, for example, being able to accurately measure the linear (non-curved) distance and directions between two local points whose geodetic coordinates are obtained from GNSS. To transform from geodetic 3D to 2D (planar) coordinates, a map projection is used to "project" from an ellipsoidal surface to a flat (grid system) surface. Projections can be carried out in a variety of ways, but whatever the method, projecting from the ellipsoid to a flat surface cannot be done without introducing some distortion in the resultant grid coordinates. Because of this effect, there is the concept of grid scale factor in map projections. Map projections are commonly categorized into three types based upon the idea of the projection surface used: azimuthal, conic, and cylindrical. Details on these can be found in [9, 11, 14, 16] and other sources.

In map projections such as State Plane Coordinate (SPC) system [6] and Universal Transverse Mercator (UTM) system, zone and polynomial constants exist for converting between geodetic and grid plane coordinates. SPC is a Lambert Conformal Conic map projection [9, 583–586] and [14, 152, 163–164], whereas UTM is a Transverse Mercator map projection [9, 586–590] and [11, 284–289]. Example formulations are provided below:

1. **State Plane Coordinates** [9, 583–586]

   SPC conversions use a Lambert Conformal Conic (LCC) projection, the NAD83 datum (GRS 80 reference ellipsoid), and zone constants. The LCC map projection is defined by two ellipsoidal parameters (semi-major axis a and eccentricity e), grid origin (φ_0, λ_0), latitude of the north standard parallel φ_1, latitude of the south standard parallel φ_2, false easting E_0, and false northing N_b. Given geodetic coordinates (φ, λ) of a point, its grid northing (N) and easting (E) coordinates, scale factor k, and convergence angle γ are computed using:

$$\begin{aligned} E &= E_0 + R\,sin(\gamma) \\ N &= N_b + R_b - R\,cos(\gamma) \\ k &= Rn/(am) \end{aligned} \tag{3.35}$$

   where

$$R_b = \text{ mapping radius at latitude of grid origin} \tag{3.36}$$

$$R = aFt^n = \text{ radius of the projection} \tag{3.37}$$

$$\gamma = (\lambda - \lambda_0)n \tag{3.38}$$

$$n = sin\,\varphi_0 = (\ln m_1 - \ln m_2)/(\ln t_1 - \ln t_2) \tag{3.39}$$

$$F = m_1/(nt_1^n) \tag{3.40}$$

$$m = cos\,\varphi(1 - e^2 sin^2\varphi)^{1/2} \tag{3.41}$$

$$t = tan(\pi/4 - \varphi/2)/[(1 - e\,sin\,\varphi)/(1 + e\,sin\,\varphi)]^{e/2} \tag{3.42}$$

   and t_i is t at φ_i, m_i is m at φ_i, and $R_b = aFt_0^n$.

2. **UTM Coordinates** [14, 162–165]

   The UTM conversion uses a Transverse Mercator (TM) cylindrical projection, and no particular reference ellipsoid is stipulated, but WGS 84 and GRS 80 ellipsoids are commonly used. The TM projection is defined by ellipsoidal parameters (semi-major axis a and eccentricity e), grid origin (φ_0, λ_0), scale factor k_0 at the central meridian λ_0, false easting E_F, and false northing N_F. Given geodetic coordinates (φ, λ) of a point, the general formulas for the computation of its UTM grid coordinates (N, E) and scale factor k are [14, p. 163] and [21, pp. 2–6]:

$$\begin{aligned} E &= E_F + k_0 E_{TM} \\ N &= N_F + k_0 N_{TM} \\ k &= k_0 k_{TM} \end{aligned} \tag{3.43}$$

where

$$\begin{aligned} E_{TM} = \nu[A + (1 - T + C)A^3/6 \\ + (5 - 18T + T^2 + 72C - 58(e')^2)A^5/120] \end{aligned} \tag{3.44}$$

$$\begin{aligned} N_{TM} = d - d_0 + \nu tan\varphi[A^2/2 + (5 - T + 9C + 4C^2)A^4/24 \\ + (61 - 58T + T^2 + 600C - 330(e')^2)A^6/720] \end{aligned} \tag{3.45}$$

$$\begin{aligned} k_{TM} = 1 + (1 + C)A^2/2 + (5 - 4T + 42C + 13C^2 - 28(e')^2)A^4/24 \\ + (61 - 148T + 16T^2)A^6/720 \end{aligned} \tag{3.46}$$

$$T = tan^2\varphi;\ C = (e')^2 cos^2\varphi;\ A = (\lambda - \lambda_0)cos\varphi \tag{3.47}$$

$$\begin{aligned} d = a[(1 - e^2/4 - 3e^2/64 - 5e^6/256 - \ldots)\varphi \\ - (3e^2/8 + 3e^4/32 + 45e^6/1024 + \ldots)sin2\varphi \\ + (15e^4/256 + 45e^6/1024 + \ldots)sin4\varphi \\ - (35e^6/3072 + \ldots)sin6\varphi + \ldots] \end{aligned} \tag{3.48}$$

and ν is radius of curvature in the prime vertical at φ, e' is second eccentricity, d_0 is the value of d (Eq. 3.48) at φ_0, and $k_0 = 0.9996$.

3.3 Reference Surfaces and GNSS-Derived Heights

There are three important reference surfaces in geodesy: (1) topography, (2) the geometric reference surface (ellipsoid), and (3) the physical reference surface (geoid). These surfaces are illustrated in Fig. 3.6. Topography is the Earth's surface upon which the measurement points are located. Ellipsoid is the mathematical figure with known parameters to approximate the size and shape of the Earth. Geoid is a non-mathematical reference surface which closely corresponds with mean sea level as defined by gravity potential and is everywhere normal to the direction of gravity.

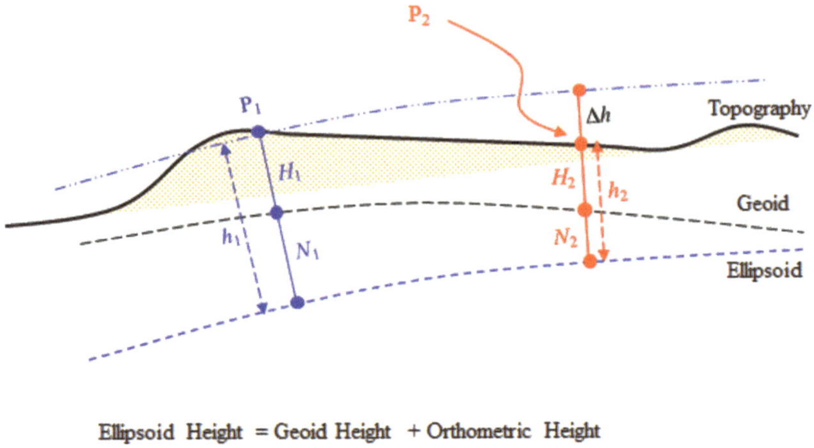

Fig. 3.6 Relationship between heights and reference surfaces

The height of a point located on the Earth's topography can be defined in three different ways: (1) the distance from that point to the ellipsoid surface is the ellipsoidal height h, (2) the distance from that point to the geoid is termed the orthometric height H, and (3) the distance between the ellipsoid and the geoid reference surfaces for that point is the geoid height N. Ellipsoidal heights are obtained from GNSS measurements but for practical purposes orthometric heights[12] (related to gravity field) are required. But since orthometric heights cannot be obtained from GNSS directly, the relation between ellipsoid height (from GNSS observations), orthometric height (from spirit leveling), and geoid height (from a geoid model or computation) can be applied. This relation can be written as follows (using the illustrations in Fig. 3.6) [16]:

$$\begin{aligned} &h_1 = N_1 + H_1;\ h_2 = N_2 + H_2, \\ &\Delta H = H_2 - H_1; \Delta h = h_2 - h_1; \Delta N = N_2 - N_1, \\ &\Delta h = \Delta H - \Delta N \end{aligned} \tag{3.49}$$

If two types of height information are known (according to Eq. 3.49), the third one can be derived. For example,

1. If precise geoid height is known (e.g., from a geoid model or geoid computation), the orthometric height can be derived from GNSS measurements (since ellip-

[12] Projects such as road constructions, pipes, and drainage systems rely on orthometric heights (elevations) which determine the direction of flow of water and liquids based on gravity potential differences.

soidal height is available from a GNSS solution) to control or substitute geodetic leveling.
2. If elevations (orthometric heights) are available from precise leveling and ellipsoidal heights are available from GNSS solutions, the geoid heights can be derived or estimated.

If only height changes are needed, then the determination of GNSS (ellipsoidal) heights without reference to the geoid will be sufficient.

3.4 Earth Orientation and Polar Motion

Satellite's ECEF coordinates (Sect. 3.2.1, Fig. 3.1) are initially derived from satellite orbits (see Appendix A) represented with a set of parameters and time. Satellite orbits and their orientation in space are defined using a space-fixed (inertial) frame, e.g., the International Celestial Reference Frame (ICRF) [7, 11, 16]. ICRF[13] is also the basis for defining the Earth's orientation and its variable rotation in space (Fig. 3.7). Understanding the relationship between terrestrial stations and orbiting satellites requires the use of Earth-fixed coordinate systems, e.g., ECEF, and a space-fixed coordinate system such as ICRF. Information on Earth orientation is required for relating the changing aspects of the Earth-based reference systems to a space-fixed system (see, e.g., [18, page 82] and [7, page 206]).

Earth orientation (depicted in Fig. 3.7) refers to the direction in space of the Earth's rotation axis, celestial equator, and the ecliptic (plane of the Earth's orbit around the sun). It is ordinarily measured using five quantities: an angle describing the rotational motion of the Earth (UT1-UTC), two angles that characterize the direction of the Earth's rotation axis in space (celestial pole), and two angles defining the direction of the Earth's rotation axis within the Earth (polar motion). With these coordinates, the orientation of the Earth in space is fully defined. The International Earth Rotation and Reference Systems Service (IERS) monitors the Earth's orientation and disseminates this information to various organizations on a continuous basis. The orientation of the Earth's rotation axis changes over time with respect to the stars. The celestial motions that cause this to happen may be explained by the physics of the dynamic system in which torques are exerted on the Earth's equatorial bulge by gravitational fields of the sun, moon, and planets. These motions are called precession and nutation (Fig. 3.8).

Precession is the slow circular movement of the Earth's rotation pole in the inertial space. Its rate is approximately 50 arc seconds annually, translating into a period of 26,000 years. **Nutation** is a more rapid motion, perturbation of the Earth's precession, which causes it to be irregular, instead of smooth. It is comprised of a

[13] The ICRF was defined and adopted by the International Astronomical Union (IAU) on January 1, 1998. It was determined using radio positions of 212 extragalactic radio sources determined through very long baseline interferometry (VLBI).

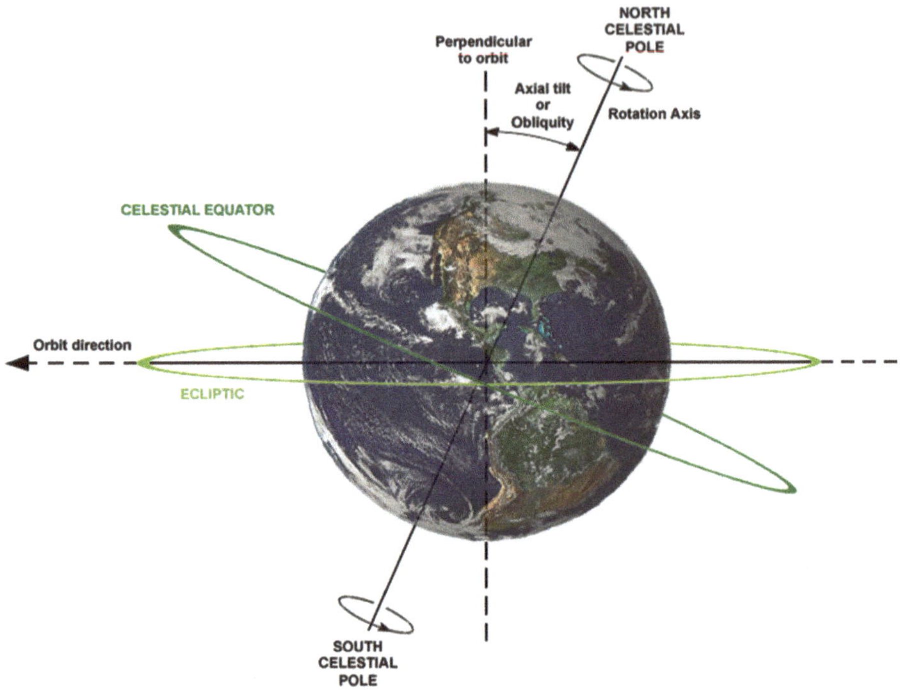

Fig. 3.7 Earth orientation frame

number of oscillations with periods ranging from 14 days to 18.6 years. Nutation is due to variations in gravitational attraction between the Earth and other celestial bodies due to constantly changing distances between them (due to their various orbits).

The angles that characterize the direction of the rotation pole within the Earth are called the polar coordinates, x and y. Variation in these coordinates is called **Polar Motion**. The polar coordinates measure the position of the Earth's instantaneous pole of rotation in a reference frame which is defined by the adopted locations of terrestrial observatories. The x coordinate is measured along the 0° (Greenwich) meridian, while the y coordinate is measured along the 90 °W meridian. These two coordinates determine the directions on a plane onto which the polar motion is projected. Polar motion consists largely of two motions, an annual elliptical component and a Chandler circular component with a period of about 435 days. These two motions define the spiral motion of the Earth's pole (Fig. 3.9).

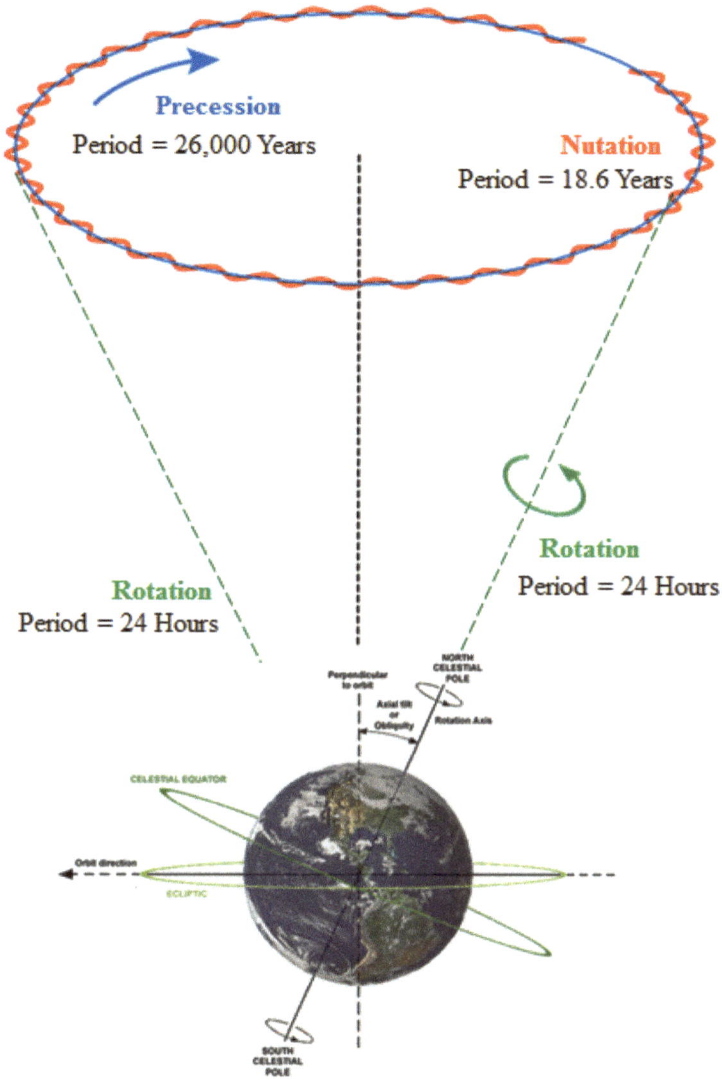

Fig. 3.8 Precession and nutation

3.5 ITRF and GNSS Reference Frames

3.5.1 International Terrestrial Reference Frame

Satellite orbits (Appendix A) are readily expressed in the ICRF, but in order to use satellites for the determination of terrestrial positions, it is necessary to transform their coordinates from the ICRF to the International Terrestrial Reference Frame

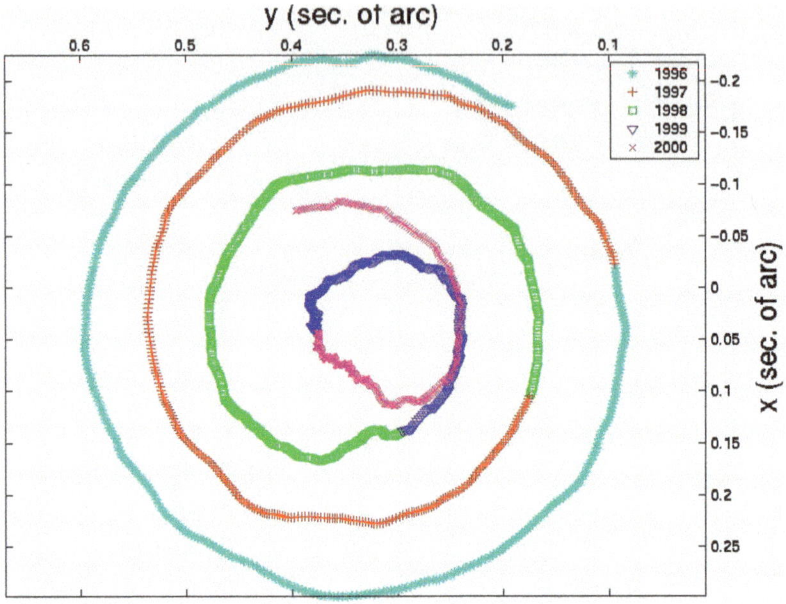

Fig. 3.9 Plot of polar motion (courtesy: U.S. naval observatory)

(ITRF). The transformation incorporates the Earth's orientation as described in the previous section. The following equation summarizes such a transformation for a given epoch t (e.g., as found in [19, pp. 43–45]):

$$\left[\mathbf{ITRF}\right] = \mathbf{R_M}(t)\mathbf{R_S}(t)\mathbf{N}(t)\mathbf{P}(t)\left[\mathbf{ICRF}\right] \tag{3.50}$$

where

$\left[\mathbf{ITRF}\right]$ is the coordinate vector in the ITRF.

$\left[\mathbf{ICRF}\right]$ is the coordinate vector in the ICRF.

$\mathbf{P}$ is the transformation matrix associated with the precession between the reference epoch and the epoch t.

$\mathbf{N}$ is the transformation matrix associated with nutation at epoch t.

$\mathbf{R_S}$ is the transformation matrix associated with the Earth's rotation around the Celestial Ephemeris Pole (CEP) axis.

$\mathbf{R_M}$ is the transformation matrix associated with the polar motion.

ICRF is the practical implementation of the *Conventional Celestial Reference System* (CRS), and ITRF is the practical implementation of the *Conventional Terrestrial Reference System* (TRS) [7, 18, 19]. Each of the two systems is defined using internationally adopted conventions as follows:

1. CRS, an inertial reference system, has its origin at the Earth's center of mass (geocenter). Its X-axis points in the direction of the mean vernal equinox at the J2000.0 epoch. The Z-axis is orthogonal to the plane defined by the mean equator at the J2000.0 epoch. The Y-axis is orthogonal to both the X-axis and the Z-axis and completes a right-handed system. The ICRF (an implementation of the CRS) is determined from a set of precise coordinates of intergalactic radio sources. Thus, the ICRF is fixed with respect to distant stars and objects of the universe.
2. TRS is an ECEF reference system (see, e.g., Fig. 3.1) that co-rotates with the Earth in its diurnal rotation. It also has its origin at the Earth's center of mass (geocenter). The Z-axis is coincident with the Earth's rotation axis (of the Earth's diurnal rotation), the X-axis lies in the equatorial plane with its positive end intersecting the Greenwich meridian, and Y-axis is orthogonal to both X-axis and Z-axis with its positive end intersecting the Earth's ellipsoid at 90° East longitude.

ITRF[14] is the practical realization of the ITRS, determined through a set of precise coordinates of points distributed on the Earth's surface, serving as reference (datum) points. The ITRF was introduced by the IERS (International Earth Rotation and Reference Systems Service) in 1992 and is continuously updated (every year from 1992 to 1997 and every few years since ITRF2000) using the latest mathematical models. The data and precise coordinates of the points contributing to the ITRF are established using multiple geodetic techniques such as GNSS, VLBI, SLR, and DORIS.

Other terrestrial reference frames that are based on the TRS include, for example, the World Geodetic System 84 (WGS-84) for the GPS constellation, the Parametry Zemli 1990 (Parameters of the Earth 1990) (PZ-90) for GLONASS, the Galileo Terrestrial Reference Frame (GTRF) for Galileo, and the China Geodetic Coordinate System 2000 (CGCS2000) for Beidou system. These are briefly described in the next section.

3.5.2 *GNSS Reference Frames*

3.5.2.1 World Geodetic System 1984 (WGS-84)

GPS satellites broadcast their orbits in WGS-84, a terrestrial reference system developed by the US Department of Defense for the GPS constellation. GPS broadcast ephemerides are linked to the position of the satellite antenna phase center in the WGS-84 ECEF reference frame. Therefore, the GPS receiver position

[14] What is the ellipsoid associated with the realizations of ITRS? The ITRF solutions do not directly use an ellipsoid. ITRF solutions are specified by Cartesian equatorial coordinates X, Y, and Z. If needed, they can be transformed to geographical coordinates (Longitude, Latitude, and Height) referred to an ellipsoid. GRS80 ellipsoid is recommended. (http://itrf.ensg.ign.fr/).

Table 3.2 Main ellipsoidal parameters of WGS-84

Parameter	Symbol	Value
Semi-major axis	a	6 378 137.0 m
Flattening factor	f	1/298.257 223 563
Earth's angular velocity	ω_E	$7\,292\,115.0 \times 10^{-11}$ rad s^{-1}
Geocentric gravitational constant	GM	$3\,986\,004.418 \times 10^{8}$ m^{3}s^{-2}
Speed of light in vacuum	c	$2.997\,924\,58 \times 10^{8}$ m s^{-1}
Second zonal harmonic	J_2^0	$-484.16685 \times 10^{-6}$

Table 3.3 Main ellipsoidal parameters of PZ-90

Parameter	Symbol	Value
Semi-major axis	a	6 378 136.0 m
Flattening factor	f	1/298.257 839 303
Earth's angular velocity	ω_E	$7\,292\,115.0 \times 10^{-11}$ rad s^{-1}
Geocentric gravitational constant	GM	$3\,986\,004.4 \times 10^{8}$ m^{3}s^{-2}
Speed of light in vacuum	c	$2.997\,924\,58 \times 10^{8}$ m s^{-1}
Second zonal harmonic	J_2^0	$1082625.75 \times 10^{-9}$

coordinates will be expressed in the same WGS-84 ECEF reference frame. The major parameters of the associated ellipsoid of WGS-84 are given in Table 3.2. The refined frame WGS-84(G1150), introduced in 2002, agrees with ITRF at the centimeter level and is highly accepted as a primary satellite-based reference coordinate system.

3.5.2.2 Parametry Zemli 1990 (PZ-90)

GLONASS satellites broadcast their orbits (ephemerides) in the PZ-90 reference frame. PZ-90 is an ECEF frame just like the WGS-84 and its associated ellipsoidal parameters as shown in Table 3.3 [12].

PZ-90.11 is the updated version of PZ-90, implemented by the GLONASS ephemeris information since 2013. The transformation from PZ-90.11 to the ITRF2008 frame applies only an origin shift, but no rotations or scale factor. The following equation shows the transformation [19]:

$$\begin{bmatrix} x \\ y \\ z \end{bmatrix}_{\text{ITRF2008}} = \begin{bmatrix} x \\ y \\ z \end{bmatrix}_{\text{PZ-90.11}} + \begin{bmatrix} 0.003 \pm 0.002\ \text{m} \\ 0.001 \pm 0.002\ \text{m} \\ 0.001 \pm 0.002\ \text{m} \end{bmatrix} \tag{3.51}$$

Table 3.4 Main ellipsoidal parameters of CGCS2000

Parameter	Symbol	Value
Semi-major axis	a	6 378 137.0 m
Flattening factor	f	1/298.257 222 101
Earth's angular velocity	ω_E	$7\ 292\ 115.0\times10^{-11}$ rad s^{-1}
Geocentric gravitational constant	GM	$3\ 986\ 004.418\times10^{8}$ m^{3}s^{-2}
Speed of light in vacuum	c	$2.997\ 924\ 58\ \times10^{8}$ m s^{-1}

3.5.2.3 Galileo Terrestrial Reference Frame (GTRF)

Galileo satellites broadcast their orbits in Galileo Terrestrial Reference Frame (GTRF). GTRF was developed by a consortium called the Galileo Geodetic Service Provider (GGSP) under the leadership of GFZ Potsdam [19]. The initial coordinates for the GTRF reference stations were provided using GPS observations, although subsequent versions have used both GPS and Galileo observations. According to Galileo requirements, the difference between the GTRF coordinates compared to the most recent ITRF should not exceed 3 cm (at the 2σ level). The first GTRF was released in 2007, and each subsequent realization incorporates more data, better computational techniques, a better knowledge of the Earth, and improved accuracy.

3.5.2.4 China Terrestrial Reference Frame (CTRF)

The Beidou system (BDS) uses CGCS2000,[15] an Earth-centered, Earth-fixed terrestrial reference system and geodetic datum associated with the Earth's ellipsoid (Table 3.4). CTRF2000 is a realization of CGCS2000 using a national network of GPS control stations and astro-geodetic observations. The definition of CTRF2000 is such that its origin has a zero translation and translation rate with respect to ITRF1997, its scale has zero scale and scale rate with respect to ITRF1997, and its orientation has zero rotation and rotation rate with respect to ITRF1997. Further realizations and updates will utilize GNSS continuously operating reference stations (CORSs).

References

1. BDS-SIS-ICD-2.0 20131226. (2013). BeiDou navigation satellite system signal in space interface control document open service signal (Version 2.0). China Satellite Navigation Office.
2. Allan, D., Ashby, N., & Hodge, C. (1997). The science of timekeeping. Hewlett Packard (HP) Application Note 1289. 88pp.

[15] China Geodetic Coordinate System 2000 (CGCS 2000) (http://www.unoosa.org/pdf/icg/2012/template/CTRF2000.pdf).

3. Aoki, S., Guinot, B., Kaplan, G. H., Kinoshita, H., McCarthy, D. D., & Seidelmann, P. K. (1982). The new definition of universal time. *Astronomy and Astrophysics, 105*(2), 359–361.
4. BIPM. (2006). SI brochure: The international system of units (SI) [8th edition, 2006; updated in 2014]. https://www.bipm.org/en/publications/si-brochure/
5. Altamimi, Z., Sillard, P., & Boucher, C. (2002). ITRF2000: A new release of the international terrestrial reference frame for earth science applications. *Journal of Geophysical Research, 107*(B10), 2214. https://doi.org/10.1029/2001JB000561
6. Dennis, M. (2018). The state plane coordinate system: History, policy, and future directions. NOAA Special Publication NOS NGS 13.
7. Elithorp, J., & Findorff, D. (2009). *Geodesy for geomatics and GIS professionals* (2nd edn.). Copley Custom Textbooks. Ann Arbor: XanEdu Publishing Inc. ISBN 1-58152-658-X.
8. Galileo-OS-SIS-ICD. (2016). European GNSS (Galileo) open service, signal-in-space interface control document. European Union. Issue 1.3.
9. Ghilani, C., & Wolf, P. (2006). *Adjustment computations: Spatial data analysis* (4th edn.). New York: Wiley. ISBN 13 978-0-471-69728-2.
10. Harvey, B. (2009). Practical least squares and statistics for surveyors. University of New South Wales (UNSW), Sydney.
11. Hofmann-Wellenhof, B., Lichtenegger, H., & Wasle, E. (2008). *GNSS - global navigation satellite systems: GPS, GLONASS, Galileo, and more*. Wien: Springer.
12. ICD GLONASS CDMA General Edition 1.0. (2016). Interface control document. General description of code division multiple access signal system (Edition 1.0), Russian Space Systems, JSC.
13. ICD-GPS-870C: NAVSTAR Next Generation GPS Operational Control Segment (OCX) to User Support Community Interface [5 June 2018]
14. Meyer, T. (2010). *Introduction to geometrical and physical geodesy: Foundations of geomatics*. Redlands: ESRI Press.
15. Powers, P. (2017). CGSIC GPS week roll over issue, US Naval Observatory (USNO), September 26. https://www.gps.gov/cgsic/meetings/2017/powers.pdf
16. Seeber, G. (2003). *Satellite geodesy*. New York: Walter de Gruyter.
17. Soler, T., & Marshall, J. (2003). A note on frame transformations with applications to geodetic datums. *GPS Solutions, 7*(2), 148–149.
18. Stolz, A. (2001). An introduction to geodesy. Monograph 16. The University of New South Wales (UNSW), Sydney.
19. Subirana, J.S., Zornoza, J.M., & Hernández-Pajares, M. (2013). GNSS Data Processing Volume I: Fundamentals and Algorithms. ESA TM-23/1.
20. The united nations office for outer space affairs (UNOOSA) BeiDou time scale description. http://www.unoosa.org/pdf/icg/2016/Beidou-Timescale2016.pdf
21. US Defense Mapping Agency. (1989). The universal grids: Universal transverse mercator (UTM) and universal polar stereographic (UPS). Defense Mapping Agency Technical Manual 8358.2.

Chapter 4
Estimating Geodetic Parameters from GNSS

4.1 Estimating Geodetic Parameters

The geodetic parameters of interest include, but are not limited to, station position coordinates and velocities; satellite position coordinates and velocities; baseline vectors; heights and height systems; the Earth's orientation and its geocenter, the geoid, and other related quantities from gravity field measurements. These parameters are estimated using precise measurements and the exact science of geodesy, for which GNSS is one of the widespread, globally accessible tools. The basic geometry for estimation of geodetic parameters from Earth-orbiting GNSS satellites is illustrated in Fig. 4.1.

GNSS satellites, like all other Earth-orbiting satellites, are kept in orbit by the Earth's gravitational forces acting upon them; hence, their orbits are centered at the Earth's geocenter. As already shown in Sect. 3.2.1, this consideration leads to the definition of a geocentric ECEF Cartesian coordinate system in which the satellite position coordinates are expressed. Thus points on, below, or above the Earth surface, whose exact positions are derived from observations of and from GNSS satellites, are primarily expressed in a geocentric coordinate system such as ECEF Cartesian or ellipsoidal system.

Since the definition of geodesy includes measuring and understanding the Earth's geometric shape, orientation in space, and gravitational field, a determination of the geocenter is vital to that definition. The geocenter is the center point of the Earth's mass (or figure) whose geometric shape and size is a fundamental property in geodesy. Mathematically, the Earth's figure is defined from zonal harmonics of the Earth's gravity field (e.g., [6]), for example, the first harmonic (spherical), second harmonic (ellipsoidal), and in a greatly oversimplified way, an Nth harmonic (where N is very large) approximate the geoid, whose center point is the geocenter. The Earth's gravity field and its variations over the Earth's surface and the complex shape of the geoid may be built up from spherical harmonic coefficients obtained from satellite methods. Perturbations to the orbital elements that are commonly used to

C. A. Ogaja, *Introduction to GNSS Geodesy*,
https://doi.org/10.1007/978-3-030-91821-7_4

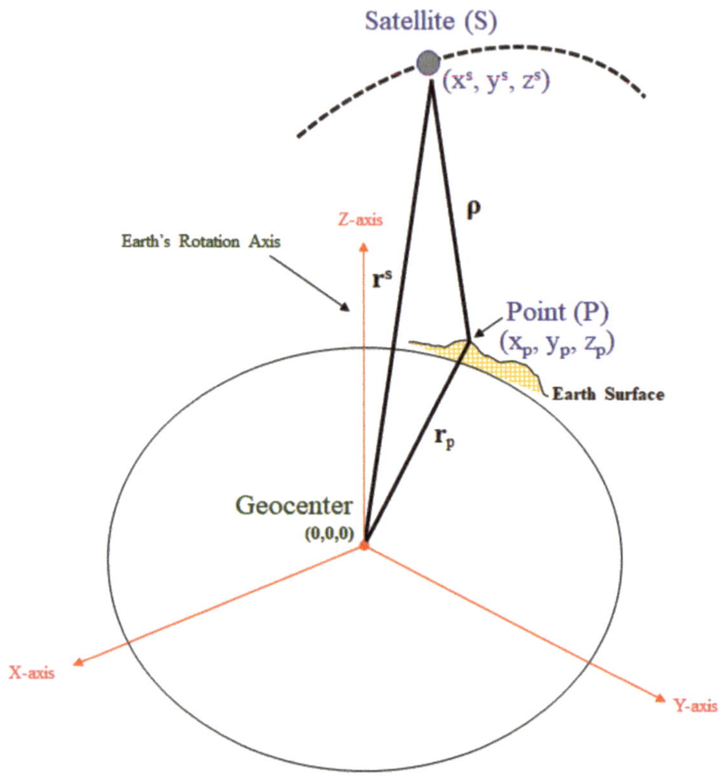

Fig. 4.1 Geometry of Satellite Observations

describe the geometry of satellite orbits (Appendix A) may also be used to determine the Earth's shape.

In Fig. 4.1, the observation vector $\boldsymbol{\rho}$ relates to the geocentric positions, $\boldsymbol{r}^s$ of the satellite and $\boldsymbol{r}_p$ of the point, according to

$$\mathbf{r}^{\mathrm{s}}(t) = \mathbf{r}_{\mathbf{p}}(t) + \boldsymbol{\rho}(t) \tag{4.1}$$

The geocentric satellite position vector, $\boldsymbol{r}^s(t)$, is a function of the orbital elements at some reference epoch, t_0, and the changes in these elements due to various perturbing forces in the interval t–t_0. The vector $\boldsymbol{\rho}$ is only partially observed, for example, an estimate of its magnitude ρ (called pseudorange) is recorded in RINEX data files (Sect. 6.3.1). The vector $\boldsymbol{r}_p$ is considered as unknown and generally time-dependent (e.g., with velocities due to crustal motion/deformation/tectonics) in the case of satellite tracking over a period of time, for example, spanning days or longer.

While oversimplified, Eq. (4.1) describes a nonlinear relation between observed quantities and unknown parameters. Solutions are obtained by observing satellites from stations that are geographically well distributed. A tracking network determines the satellite orbits, i.e., the parameters defining $\boldsymbol{r}^s(t)$, by tracking satellites over a period of time continuously. The estimations of satellite orbits, $\boldsymbol{r}^s(t)$, and station positions, $\boldsymbol{r}_p(t)$, are in fact interdependent. For example, a tracking network whose positions are known determines the satellite orbits by tracking satellites over a period of time, and new points or station positions are subsequently established by tracking satellites whose orbits, or orbit prediction models, are known.

GNSS satellites are observed continuously from a global network of permanent tracking stations (Chap. 6), and their orbits are computed at regular and frequent intervals. The satellite positions and velocities are extrapolated ahead of time, and this information is transmitted from the satellite at the time of observation. Such orbits are accurate to a few meters. More accurate orbits may be obtained from additional GNSS data taken simultaneously from sites whose positions are well known from independent measurements such as VLBI[1] or SLR.[2]

In most applications where, for example, the goal is to establish new points or station positions (the vector $\boldsymbol{r}_p(t)$ in Eq. (4.1)), simultaneous observations at two or more stations are considered, whereby some of the unknown parameters and systematic biases are eliminated in the observation equations by mathematically differencing the measurements (Sect. 5.3.1), leaving only the desired parameters such as the unknown station positions and a few measurement biases. Such methods produce relative positions, but if absolute geocentric coordinates of one or more network (base) stations are known, the solution can be centered at the geocenter.

4.2 Mathematical Concept

4.2.1 Estimation by Least Squares

The classical approach in the estimation of geodetic parameters is through the least squares method [1]:

$$\tilde{y} = Ax + v; \quad v \sim N(0, \sigma_0^2 W^{-1}) \tag{4.2}$$

where $\tilde{y}$ is an $m \times 1$ vector of observations, A is an $m \times n$ design matrix, x is an $n \times 1$ vector of unknown parameters, v is an $m \times 1$ vector of observational errors, σ_0^2 is the variance of unit weight, W is an $m \times m$ positive-definite weight matrix, and $m \geq n$ for a least squares solution.

[1] Very Long Baseline Interferometry (VLBI) is a type of astronomical interferometry used in radio astronomy. It is uniquely suited for high-precision global geodesy and things such as reference frame scale definition.

[2] Satellite Laser Ranging (SLR) is a method to measure distance to Earth-orbiting satellites using a powerful laser. It is uniquely suited for accurate determination of the geocenter and satellite orbit parameters.

The estimation model in Eq. (4.2) is based on the assumption that the observational errors in the vector v are random and normally distributed. The parameter estimates may then be obtained as follows (under the least squares condition that $v^T W v$ be minimum):

$$
\begin{aligned}
\hat{x} &= (A^T W A)^{-1} A^T W \tilde{y} \\
D(\hat{x}) &= \hat{\sigma}_0^2 (A^T W A)^{-1} \\
\hat{v} &= \tilde{y} - A\hat{x} \\
\hat{\sigma}_0^2 &= \hat{v}^T W \hat{v}/(m-n)
\end{aligned}
\tag{4.3}
$$

where $D(\hat{x})$ is the dispersion matrix, also known as a covariance matrix, variance matrix, or variance–covariance matrix, of the estimated quantities $\hat{x}$ and $\hat{\sigma}_0^2$ is their a posteriori variance.

The above model (Eq. (4.2)) describes a linear relation between observations and unknown parameters. However, Eq. (4.1) is a nonlinear relation in which GNSS observations involve measuring/observing the range (distance) ρ from the satellite to point of interest whose unknown (position) coordinates (in $\boldsymbol{r_p}$) are the unknown geodetic parameters.

When distance, or range, is expressed in terms of position coordinates, in a three -dimensional space, the equation is nonlinear. For example, an observation equation expressing the observed quantity ρ in terms of the satellite positions $\boldsymbol{r}^s$ and the unknown parameters $\boldsymbol{r_p}$ is defined as

$$
\rho(t) = |\boldsymbol{\rho}(t)| = |\mathbf{r_p}(t) - \mathbf{r}^{\mathrm{s}}(t)| \tag{4.4}
$$

Later we return to this nonlinear observation equation. First, we address the model in Eq. (4.2) to clarify a few things.

4.2.1.1 Redundancy

In a least squares solution (Eq. (4.2)), for n unknown parameters in vector x, at least $m = n$ observations in vector y are needed. However in practice, more measurements than necessary are typically taken, and thus $m \geq n$. The excess of m over n is referred to as redundancy. There are three unknown parameters in the position vector $\boldsymbol{r_p}$, and as illustrated later in Chap. 5, a fourth unknown (the clock bias) is usually introduced. Each of the unknown parameters requires an independent measurement. Therefore, a minimum of four range observations (that is, from four different satellites at the point of interest) are needed for a unique solution. For each of the range observations, an observation equation (Eq. (4.4)) is formulated, to express the observed quantity in terms of the four parameters of interest.

4.2.1.2 Minimum Variance Estimator

In Eq. (4.3), the weight matrix W is interchangeable with the variance–covariance matrix of the observables Q_{yy}, such that the equation can also be written as

$$\hat{x} = (A^T Q_{yy}^{-1} A)^{-1} A^T Q_{yy}^{-1} \tilde{y} \tag{4.5}$$

This equation implies that observables with small variances should get more weight in the solution than observables with larger variances. Thus, the former are more precise than the latter. The variance–covariance matrix Q_{yy} reduces to a scaled (identity) matrix I_{yy} if all variances or weights are assigned equally to all the observables.

The above estimator has three distinct properties [8]: First, $\hat{x}$ is a *linear* function of the observables $\tilde{y}$; second, the estimator is *unbiased*; and thirdly, the estimators in $\hat{x}$ have *minimum variance*. The uncertainty of the entries in vector $\hat{x}$ is obtained from the $n \times n$ variance matrix:

$$Q_{\hat{x}\hat{x}} = (A^T Q_{yy}^{-1} A)^{-1} \tag{4.6}$$

It can be shown that the matrix in Eq. (4.6) has minimum trace (that is, the estimator is best in the sense that the sum of all n variances together is the smallest). Minimum variance implies best accuracy, and with this property, the estimator in Eq. (4.5) is also referred to as the Best Linear Unbiased Estimator (BLUE). It provides a generalization of the least squares estimation in which one can compute a properly weighted least squares solution (for vector x) to any proper linear problem.

4.2.1.3 Nonlinear Observation Equations

A linear model of observation equations has been presented in Eq. (4.2), from which the least squares estimate, the Best Linear Unbiased Estimator (BLUE), is shown in Eq. (4.5). This provides a nice theory but in practice not all measurements have a linear relation with unknown parameters. For example, a distance (or range) between two points (such as between a satellite in space and an Earth-based point or station) is a nonlinear function of the coordinate differences (Eq. (4.4)).

The approach to systems of nonlinear observation equations is to approximate them by linear equations. The nonlinear equations are linearized with respect to the unknown parameters, and the resulting system of linear(ized) equations is handled using Eq. (4.3) or (4.5) or (4.2).

The nonlinear model of observation equations is

$$\tilde{y} = F(x) \tag{4.7}$$

where the matrix–vector product Ax (in Eq. (4.2)) has been replaced by a nonlinear function $F(x)$, which is a collection of m nonlinear functions of n parameters. Thus, $F(x)$ can be expanded as

$$\begin{pmatrix} \tilde{y}_1 \\ \tilde{y}_2 \\ \vdots \\ \tilde{y}_m \end{pmatrix} = \begin{pmatrix} f_1(x_1, x_2, \cdots, x_n) \\ f_2(x_1, x_2, \cdots, x_n) \\ \vdots \\ f_m(x_1, x_2, \cdots, x_n) \end{pmatrix} \tag{4.8}$$

4.2.1.4 Linearization

The function $F(x)$ is approximated by the zero-order and first-order terms of Taylor Series expansion [3] in which the higher order terms are neglected. The approximation uses a vector $x_o = (x_{1(o)}, x_{2(o)}, \ldots, x_{n(o)})^T$ that contains the approximate values for all the n unknown parameters (the approximate values are presumed to be reasonably close to the actual/true x). Thus:

$$F(x) \approx F(x_o) + \left.\frac{\partial F(x)}{\partial x^T}\right|_{x_o} (x - x_o) \tag{4.9}$$

The zero-order term and the first-order derivative are evaluated at x_o. All the m nonlinear functions are each differentiated with respect to $x_1, x_2, \ldots x_n$, to compute the first-order derivative, hence creating a $m \times n$ matrix. The first row of the matrix consists of $\frac{\partial f_1}{\partial x_1} \frac{\partial f_1}{\partial x_2} \cdots \frac{\partial f_1}{\partial x_n}$, with the partial derivatives evaluated at x_o. Substituting the approximation of $F(x)$ into the model Eq. (4.7) produces

$$\tilde{y} \approx F(x_o) + \left.\frac{\partial F(x)}{\partial x^T}\right|_{x_o} (x - x_o);\; D(\tilde{y}) = Q_{yy} \tag{4.10}$$

or

$$\underbrace{\tilde{y} - F(x_o)}_{\Delta\tilde{y}} \approx \underbrace{\left.\frac{\partial F(x)}{\partial x^T}\right|_{x_o}}_{A} \underbrace{(x - x_o)}_{\Delta x};\; D(\tilde{y} - F(x_o)) = Q_{yy} \tag{4.11}$$

The $m \times n$ matrix of first-order derivatives takes the role of the design matrix A. The observations vector $\tilde{y}$ is replaced by $\tilde{y} - F(x_o)$, the observations minus their approximates based on the approximate value x_o of the unknown parameters (i.e., $y_o = F(x_o)$). Subsequently, the least squares will not be estimating the vector x of unknown parameters, but instead the differences of x with respect to approximate values x_o: $\Delta x = x - x_o$.

4.2.1.5 Estimation

Adopting the approximation model in Eq. (4.11), the estimator for Δx becomes

$$\Delta\hat{x} = (A^T Q_{yy}^{-1} A)^{-1} A^T Q_{yy}^{-1} (\tilde{y} - F(x_o)) \tag{4.12}$$

and

$$Q_{\Delta\hat{x}\Delta\hat{x}} = (A^T Q_{yy}^{-1} A)^{-1}$$

from which the estimator for the vector of the unknown parameters is obtained as $\hat{x} = x_o + \Delta\hat{x}$; and $Q_{\hat{x}\hat{x}} = Q_{\Delta\hat{x}\Delta\hat{x}}$.

The model in Eq. (4.9) is only an approximation of the actual nonlinear model. For the approximation to be good and valid, the approximate value x_o should be close to the true unknown value x. Therefore, the above procedure is repeated (iterated), starting off with as good as possible guess for x_o. Next the estimate $\hat{x}$ is determined and taken as a new approximate value that is likely closer to the true, unknown x than x_o was, and the process is repeated until an acceptable convergence is achieved. This iterative procedure is known as the Gauss–Newton method.

4.2.2 *Reducing Errors and Biases*

All measurements encounter errors and biases. The observed distance (range) between the satellite in space and an Earth-based point or station (Eq. (4.4)) is governed by the physics of GNSS radio waves as they travel through the different layers of the Earth's atmosphere. Thus it contains inherent observational errors and systematic biases such as the effects of the different layers of the atmosphere on the radio waves. Steps are taken to ensure that any such errors and/or biases encountered in the measurement process are eliminated, or reduced as much as possible, to achieve the best estimate possible, of the geodetic parameters of interest (Sect. 4.2.1.5).

A detailed discussion of the GNSS signal propagation errors and biases, as well as the various techniques and methods to reduce them, is presented in Chap. 5 (Sects. 5.2 and 5.3). Once the errors and biases are reduced, or eliminated, the remaining observational errors are assumed to be random and normally distributed, making the estimation model (in Eqs. (4.3) and (4.12)) robust and valid. It is equally noted that in some special cases, some of the systematic biases can be modeled into the estimation equation as part of the unknown parameters. For example, the estimation of atmospheric effects (troposphere, ionosphere) and satellite and/or receiver clock biases may be of interest in some applications as will be seen in later chapters.

4.2.3 Combining Multiple Data

The foundational GNSS data for geodesy comprises: (a) raw data files containing the satellite observables (time-tagged pseudoranges, carrier phases, orbit information, and so forth) and (b) the resulting solutions and error estimates from processing of the observables. The latter is the subject of this section. Ordinarily, normal equations, e.g., from multiple consecutive single solutions of station coordinates (e.g., hourly or daily solutions), can be combined into a single model of multi-hour, multi-day, or weekly solutions [5]. Similarly, reprocessing of multi-year data to benefit from long duration and improvements in models (for better understanding of plate tectonics and reference frames) can help determine improved coordinates and secular velocities for continuously operating reference stations (Chap. 6).

In Eq. (4.2), the A matrix (the design matrix) is also known as the observation matrix. However, $A^T WA$ in Eq. (4.3) is called a normal matrix, and in fact, Eq. (4.3) is the solution of a *normal equation*. This is similarly the case for Eqs. (4.5) and (4.12).

Matrices $A^T WA$ and $A^T W\tilde{y}$ in Eq. (4.3) are together called the *normal equation* of the solution and can be stored for later use in a reprocessing to regenerate results without having to repeat the steps of starting off with the raw observations. This similarly applies to the corresponding matrices in Eqs. (4.5) and (4.12). For instance, the normal equation from processing a 1-hour RINEX file (Chap. 6, Sect.6.3.1) can be saved and used in combination with normal equations from other 1-hour files in order to generate a combined solution for GNSS station coordinates.

The following is an example description of the steps on how to generate a matrix combination of normal equations:

First, the input normal equations, which are an output from another program (or same program in a previous run), are expressed in a given matrix format (e.g., based upon some conventional guidelines). For example, assume a matrix structure that includes the upper triangular portion of a symmetric matrix of coefficients of $A^T WA$, augmented by the vector $A^T W\tilde{y}$, parameter starting values, and parameter labels corresponding to the rows and columns of the normal equation coefficients matrix [4].

For matrix combination, each of the input matrix structures, including the upper triangular portion of coefficients matrix, the parameter list, the vector, and the parameter nominal values, is expanded to the full rank of the number of unique parameters among the input matrices. The rows and columns of the coefficients matrix are kept in order by ascending parameter label value. The rows and columns used to expand the matrix are filled with zeros.

The expanded matrices are then added. This method of combining normal equations preserves the symmetry so that only the triangular upper portion of the input and resulting matrices are required.

A simple illustration is included here next.

$$NEQ(1): \begin{bmatrix} 1 & 2 & 3 \\ \cdot & 4 & 5 \\ \cdot & \cdot & 6 \end{bmatrix} \begin{bmatrix} \tilde{x}_a \\ \tilde{x}_b \\ \tilde{x}_c \end{bmatrix} = \begin{bmatrix} 7 \\ 8 \\ 9 \end{bmatrix} \quad (4.13)$$

$$NEQ(2): \begin{bmatrix} 6 & 5 & 4 \\ \cdot & 3 & 2 \\ \cdot & \cdot & 1 \end{bmatrix} \begin{bmatrix} \tilde{x}_a \\ \tilde{x}_c \\ \tilde{x}_d \end{bmatrix} = \begin{bmatrix} 7 \\ 8 \\ 9 \end{bmatrix} \quad (4.14)$$

$$NEQ(1): \begin{bmatrix} 1 & 2 & 3 & 0 \\ \cdot & 4 & 5 & 0 \\ \cdot & \cdot & 6 & 0 \\ \cdot & \cdot & \cdot & 0 \end{bmatrix} \begin{bmatrix} \tilde{x}_a \\ \tilde{x}_b \\ \tilde{x}_c \\ \tilde{x}_d \end{bmatrix} = \begin{bmatrix} 7 \\ 8 \\ 9 \\ 0 \end{bmatrix} (Expanded) \quad (4.15)$$

$$NEQ(2): \begin{bmatrix} 6 & 0 & 5 & 4 \\ \cdot & 0 & 0 & 0 \\ \cdot & \cdot & 3 & 2 \\ \cdot & \cdot & \cdot & 1 \end{bmatrix} \begin{bmatrix} \tilde{x}_a \\ \tilde{x}_b \\ \tilde{x}_c \\ \tilde{x}_d \end{bmatrix} = \begin{bmatrix} 7 \\ 0 \\ 8 \\ 9 \end{bmatrix} (Expanded) \quad (4.16)$$

$$NEQ(c): \begin{bmatrix} 7 & 2 & 8 & 4 \\ \cdot & 4 & 5 & 0 \\ \cdot & \cdot & 9 & 2 \\ \cdot & \cdot & \cdot & 1 \end{bmatrix} \begin{bmatrix} \tilde{x}_a \\ \tilde{x}_b \\ \tilde{x}_c \\ \tilde{x}_d \end{bmatrix} = \begin{bmatrix} 14 \\ 8 \\ 17 \\ 9 \end{bmatrix} (Combined) \quad (4.17)$$

Even though each of the two normal equations has three parameters, between them there are four unique parameters $\tilde{x}_a$, $\tilde{x}_b$, $\tilde{x}_c$, and $\tilde{x}_d$. *NEQ*(1) is expanded to add a zero-filled row and column for $\tilde{x}_d$, and *NEQ*(2) is expanded with a zero-filled row and column for $\tilde{x}_b$. The expanded matrices and right-hand side vectors are then added element by element to form the combined normal equation (Eq. (4.17)).

The normal equation model is also especially useful for combining multi-technique data (e.g., from GPS/GNSS, SLR, VLBI, DORIS, gravimetry, and leveling) that is outside the scope of this book. Such multi-technique combinations are based on the combined normal equation systems for estimating parameters such as station coordinates, Earth orientation, and troposphere parameters. For further reading, see for example [2] and [7].

References

1. Aduol, F. (2003). Robust geodetic parameter estimation under least squares through weighting on the basis of the mean square error. In E. W. Grafarend, F. W. Krumm, & V. S. Schwarze (Eds.), *Geodesy-The challenge of the 3rd millennium*. Berlin: Springer.
2. Altamimi, Z., Sillard, P., & Boucher, C. (2006). CATREF software: Combination and analysis of terrestrial reference frames.
3. Hazewinkel, M. (Ed.) (2001). *[1994], "Taylor series", Encyclopedia of mathematics*. Springer Science+Business Media B.V./Kluwer Academic Publishers. ISBN 978-1-55608-010-4.
4. NASA (1992) SOLVE Mathematical Formulation.
5. Steigenberger, P. (2009). *Reprocessing of a global GPS network*. Munchen: Deutsche Geodatische Kommission. ISBN 978-3-7696-5052-5.
6. Stolz, A. (2001). *An introduction to geodesy*. Monograph 16. Sydney, Australia: The University of New South Wales (UNSW).
7. Thaller, D. (2008). *Inter-technique Combination Based on Homogeneous Normal Equation Systems Including Station Coordinates, Earth Orientation and Troposphere Parameters*. Dissertation, Technische Universität München.
8. Tiberius, C. (2019). Primer on Mathematical Geodesy, CTB3310/CTB3425. Delft University of Technology.

Chapter 5
GNSS Observation Models

5.1 Geometric Range Modeling

GNSS observations are one-way ranges deduced from the comparison of a signal generated and transmitted by a satellite with a reference signal generated by a receiver. The range observations are primarily deduced from navigation codes modulated onto a carrier signal but can be more precisely computed from the phase of the carrier wave itself.

The *travel time*[1] by a GNSS signal to propagate from the phase center of the satellite antenna to the phase center of the receiver antenna is the primary observable. It is multiplied by the *speed of light* to give the *pseudorange*, an apparent range between the satellite and the receiver, which does not match its *geometric distance* (ρ in Fig. 4.1) due to factors such as synchronization errors between the satellite and receiver clocks, signal propagation through the atmosphere (ionosphere and troposphere), relativistic effects, multipath, and receiver noise [30]. The signal *travel time* is computed from the pseudorandom noise (PRN) sequences (i.e., the *navigation code*) modulated onto a carrier signal (see, for example, [30, p. 65]).

Besides the *navigation code*, the *phase* of the carrier signal itself is used to obtain a measure of the range between the satellite and the receiver. *Carrier phase* measurements are much more precise than the code measurements (typically two orders of magnitude more precise), but they are *ambiguous* by an unknown integer number of wavelengths (λN). This ambiguity changes arbitrarily every time the receiver loses the lock on the signal, producing jumps or range discontinuities (aka cycle slips). Nevertheless, the *carrier phase* is a more important measurement for high precision.

---

[1] A GNSS receiver determines signal travel time Δt by correlating the received code from satellite with a replica of this code generated in the receiver, so this replica moves in time (Δt) until the maximum correlation is obtained.

C. A. Ogaja, *Introduction to GNSS Geodesy*,
https://doi.org/10.1007/978-3-030-91821-7_5

5.1.1 Code Pseudorange Observation Equation

The code pseudorange observation of a navigation code transmitted by a satellite at emission time t^{sat} and recorded by a receiver at reception time t_{rcv} is given, in units of length, by

$$R = c(t_{rcv} - t^{sat}) = c\Delta t \tag{5.1}$$

where R is the code pseudorange measured by the receiver, c is the speed of light (in m/s), t_{rcv} is the receiver clock reading at signal reception time, t^{sat} is the signal transmission time as given by the satellite clock, and Δt is the signal travel time (in s).

By referring the receiver and satellite clock readings (t_{rcv} and t^{sat}) to an arbitrary common time scale, and introducing corresponding clock synchronization errors, Eq. (5.1) may be written as

$$R = c((T_r + \delta t_{rcv}) - (T^{\epsilon} + \delta t^{sat})) = c\Delta T + c\delta t_{rcv} - c\delta t^{sat} \tag{5.2}$$

where T_r and T^{ϵ} are the signal reception and emission times referring to a common time scale, δt_{rcv} and δt^{sat} are the receiver and satellite clock synchronization errors (i.e., $t_{rcv} = T_r + \delta t_{rcv}$ and $t_{sat} = T^{\epsilon} + \delta t^{sat}$), and ΔT is the *clock bias*-free signal travel time.

GPS time is typically used as the common time scale, even though such an approach is most reasonable in a GPS-only analysis. In cases of system-specific applications, time should refer to the corresponding system time scale. For example, in a GLONASS-only analysis, time should refer to the GLONASS time scale. However, in multi-GNSS applications, the choice of a common reference time scale is a matter of convention [19].

The term $c\Delta T$ in Eq. (5.2) represents the geometric distance between receiver and satellite plus signal delays (e.g., due to the Earth's atmosphere) and other applicable errors. Thus, a more detailed version of the observation equation is established by representing the term $c\Delta T$ in terms of components and pertinent error terms as follows:

$$\begin{aligned} R = &|(\boldsymbol{P} + \boldsymbol{E} + \boldsymbol{O}) - (\boldsymbol{p} + \boldsymbol{e} + \boldsymbol{o})| + c\delta t_{rcv} - c\delta t^{sat} \\ &+ cB - cb + d_{trop} + d_{ion} + d_{orb} + \gamma + M + v \end{aligned} \tag{5.3}$$

where:

- $\boldsymbol{P}$ is the geocentric position vector of the *station* (receiver) *reference point* in an Earth-fixed reference frame at observation time (measurement epoch).
- $\boldsymbol{p}$ is the geocentric position vector of the satellite's center of mass (COM) in the same Earth-fixed reference frame at observation time (see Sect. 3.5.1 for the transformation between the inertial and the Earth-fixed frame).

$\boldsymbol{E}$	is the station antenna eccentricity, i.e., the vector pointing from the *station reference point* to the antenna reference point (ARP).
$\boldsymbol{e}$	is the satellite antenna eccentricity, i.e., the vector pointing from the satellite's COM to the satellite's ARP.
$\boldsymbol{O}, \boldsymbol{o}$	are the antenna phase center vectors of the station and the satellite w.r.t. their ARPs (i.e., vectors pointing from ARP to the antenna phase center).
B, b	are code biases caused by receiver and satellite hardware delays of the ranging code signal.
d_{trop}	is the signal delay due to the Earth's troposphere.
d_{ion}	is the signal delay due to the Earth's ionosphere.
d_{orb}	is the orbital (ephemeris) error.
γ	is a correction term due to relativistic effects.
M	is the code multipath caused by reflected signals.
v	are all remaining unmodeled effects and observation noise.

The assumption is that the GNSS station (receiver) is located on the Earth surface, or on a structure or platform fixed to the Earth surface (as opposed to being airborne or space-borne). Therefore, the position vector $\boldsymbol{P}$ also contains various displacements, such as due to plate tectonics, solid Earth tides, pole tides, ocean loading, and post-glacial rebound [7, 19, 22].

The antenna phase center vectors $\boldsymbol{O}$ and $\boldsymbol{o}$ are comprised of phase center offsets (PCOs) and phase center variations (PCVs). The PCO is the offset of ARP, or the antenna physical (geometrical) center, from antenna phase center (APC), i.e., antenna's *mean electromagnetic reference point.*[2] The actual position of the APC varies with the direction (elevation, azimuth) and intensity of the satellite, the frequency of the incoming signal, and is modeled by the PCVs. Further detailed information on antenna phase center modeling and calibration can be found in [4, 5, 8, 18, 26, 27].

Signal delays due to troposphere and ionosphere are part of the common error terms in the observation equation. Strategies and models for handling them are well studied and developed (see, e.g., [6, 31]), and higher order ionospheric effects are comprehensively reviewed in [24].

GNSS provide several navigation codes on different carrier waves. These are detailed in the RINEX manual [25] and [13]. In the observation equation (5.3), the satellite biases b are caused by frequency dependent satellite hardware signal delays for the individual code types. However, the receiver bias(es) B depends not only on the code type but also on the particular tracking mode and the carrier frequency. Receiver tracking modes[3] are methods that have been developed to track encrypted

[2] Antenna's mean electromagnetic reference point, also called electrical antenna phase center (APC), is the transmission or reception point of a carrier wave from satellite to receiver. APCs are unique to each antenna (hardware dependent) and are defined by the electromagnetic properties of the antennas. For example, for one antenna, the L1 APC, L2 APC, and L5 APC are all different.

[3] Some GNSS navigation codes are encrypted for security reasons. Thus, a direct tracking of these code types is not possible for unauthorized users. However, several methods (receiver tracking

codes. These modes may lead to different signal hardware delays (the bias B) in the receiver.

The relativistic clock correction term γ is due to general and special relativity [1], i.e., the rate of two identical clocks, one placed in the satellite and the other on the ground-based receiver (or on or near-Earth surface), will differ due to the difference of the gravitational potential (general relativity) and the relative speed between them (special relativity).

The multipath error M is caused by the interference of direct and reflected signals at the receiving antenna (see, e.g., [3] and [13]).

5.1.2 Phase Observation Equation

Besides the navigation code, the carrier phase is also used to obtain a measure of the apparent distance between satellite and receiver. This yields a much more precise measure than the code, but they are ambiguous by an unknown integer number of wavelengths (λN).

One of the challenges of using carrier phase is the arbitrary change of phase ambiguity every time the receiver experiences loss of lock on the signal, producing jumps (cycle slips) or range discontinuities (gaps).

The phase observation emerges from the comparison of a received carrier wave with a reconstructed one in the receiver. The difference between these carrier waves yields a fractional part and an integrated integer number of phase cycles. However, the number of integer phase cycles is initially unknown and is referred to as the phase ambiguity. The phase observation equation has similar structure as the code pseudorange observation equation (5.3), but with additional phase-related parameters [19]:

$$\begin{aligned} \Phi = &|(\boldsymbol{P} + \boldsymbol{E} + \boldsymbol{O}) - (\boldsymbol{p} + \boldsymbol{e} + \boldsymbol{o})| + c\delta t_{rcv} - c\delta t^{sat} \\ &+cB - cb + d_{trop} - d_{ion} + d_{orb} + \lambda(\phi_{rcv} - \phi^{sat}) \\ &-\lambda N + \lambda W + \gamma + M + v \end{aligned} \tag{5.4}$$

where:

B, b	are phase biases caused by receiver and satellite hardware delays of the carrier wave.
λ	is the carrier wavelength.
ϕ_{rcv}, ϕ^{sat}	are the initial phase readings of receiver and satellite at an arbitrary start epoch.
N	is the initial phase ambiguity.

modes) have been developed to circumvent the problem. These modes may lead to different signal hardware delays in the receiver.

W	is the phase polarization effect (also known as phase wind-up).
M	is the phase multipath caused by reflected signals.
v	are all remaining unmodeled effects and observation noise.

First, it is important to note the opposite signs for the term d_{ion} in Eqs. (5.3) and (5.4). This is because ionospheric refraction causes ionospheric signal delay for code observations, whereas it causes a phase advance in the carrier phase observations.

The receiver phase bias B depends on the carrier frequency but also on the tracking mode just like the receiver code bias in Eq. (5.3). However, the satellite bias b, unlike in the code pseudorange equation, only depends on the frequency of the carrier wave but not on the code type.

The initial phase readings, ϕ_{rcv}, ϕ^{sat}, of the receiver and satellite are constant as long as the equipment is not reset or restarted.

As noted earlier, the phase ambiguity N remains constant as long as there is no loss of lock on the GNSS signal. In the event of loss of lock (causing gaps or cycle slips), a new ambiguity term is set up after signal is re-acquired. Estimating the correct integer values (i.e., ambiguity resolution) and applying in the solution of Eq. (5.4) as known parameter(s) (i.e., ambiguity fixing) strengthen the solution. See, e.g., [13] and [31] for details.

The phase wind-up W applies only to carrier phase measurements and is due to electromagnetic nature of circularly polarized waves that are commonly used by GNSS [34]. A change in the orientation of the satellite's antenna with respect to the receiving antenna (i.e., a rotation[4]) causes a phase variation either as a carrier phase advance or delay.

5.2 GNSS Error Sources

When the satellite signals travel from the satellite to the receiver, the signals are affected by atmospheric delay and multipath. In addition to these error sources, there are satellite related biases (e.g., satellite orbit and satellite clock errors) and receiver related biases (receiver clock bias, internal inter-channel biases, antenna phase center variation, and receiver noise).

[4] "For a receiver with fixed coordinates, the wind-up is due to the satellite orbital motion. As the satellite moves along its orbital path, it must perform a rotation to keep its solar panels pointing to the Sun direction in order to obtain the maximum energy while the satellite antenna keeps pointing to the Earth's center. This rotation causes a phase variation that the receiver misunderstands as a range variation." [30]

5.2.1 Atmospheric Effects

The Earth's atmosphere that affects the propagation of GNSS signals consists of two main layers, the ionosphere and troposphere. The atmospheric effects can be explained, in a general context, using Snell's law[5] in physics, whereby GNSS signals have a distorted geometric path when traveling through the atmosphere due to the refractive index gradients.

As the GNSS signals leave the satellites, they first pass through the ionosphere (the band of the atmosphere from around 50 to 1000 km above the Earth's surface) [13]. It consists of layers of electrically charged particles or ions; hence, GNSS signals do not travel at the speed of light as they transit the layer. The measured pseudoranges become too long (Eq. (5.3)), while the measured phase ranges become too short (Eq. (5.4)). The ionospheric delay is a function of the total electron content (TEC) along the signal path and the frequency of the signal. The TEC depends on time, season, and geographic location, with major influencing factors being the solar activity and the geomagnetic field. In temperate zones, the ionosphere is more stable compared to polar and equatorial regions, where the ionosphere is less stable and errors can be greater. The effect caused by neglecting signal path curvature due to ionosphere can reach up to several meters at low elevations and under high solar activity conditions [14, 15].

The troposphere is that band of the atmosphere from the Earth's surface to about 50 km. The tropospheric delay is a function of elevation and altitude of the receiver and is dependent on many factors such as the atmospheric pressure, temperature, and water vapor content. The effect caused by neglecting signal path curvature due to troposphere is about 2.3 m at zenith for dry component and few centimeters for wet component and increases to higher magnitude at lower elevations [20].

5.2.2 Multipath

Multipath[6] is caused by nearby reflecting surfaces at the receiving antenna location. Just as light reflects off a shiny surface, radio signals can be reflected by solid objects and surfaces. GNSS signals cannot penetrate solid objects such as buildings, thick tree canopy, cars, ships, bridges—instead, these objects deflect them causing

[5] The density of the atmospheric layers is not homogeneous. This causes spatial and temporal variations in the refractive index. Snell's law states that $n_1 \cdot \varphi_1 = n_2 \cdot \varphi_2$ if an electromagnetic wave travels from a medium with refractive index n_1 to a second one with refractive index n_2 and crosses the distance between them at an angle φ_1 and deviates (leaves) at an angle φ_2.

[6] Since multipath error depends on the receiver's environment, to reduce its effects, (1) good data collection site(s) is necessary to avoid reflective environments, (2) use a good quality antenna that is multipath-resistant and/or can internally digitally filter out the multipath disturbance, and (3) avoid, de-weight, or minimize the use of low elevation satellite data.

the signals to arrive at the receiver via multiple paths. The reflected signals will interfere with the signals that are received via a direct path. The reflected signal travels a longer time, instead of traveling a direct signal path. As a result, this causes the receiver position to be calculated incorrectly, with the position shifting in the direction of the multipath source. Theoretically, for GPS, the maximum pseudorange multipath error is approximately one chip length of the code (that is, about 300 m for the C/A code, and approximately 30 m for the P code), while the maximum carrier phase multipath error is about a quarter of the wavelength (that is, about 5 cm for the L1 carrier, and 6 cm for the L2 carrier).

5.2.3 *Orbital Errors*

Orbital errors are due to the discrepancies from a satellite being in a different location from its predicted location.[7] These errors occur when the GNSS signal does not transmit the correct location of the satellites. They can be reduced by using disseminated orbit products such as *precise ephemeris* data, especially if the solution is not needed in real time.

In general, three types of orbit products are available for the determination of satellite position at any instant: the almanacs, broadcast ephemeris, and precise ephemeris. The first two are available in real time, while precise ephemeris, being the most accurate and only available after the fact, is used in post-processing applications. Other disseminated orbit products are also available for post-processed and real-time applications.

5.2.4 *Satellite and Receiver Clock Offsets*

The satellite and receiver clock offsets (biases) are due to clock synchronization errors referring to the GNSS time scale (see Eq. (5.2)).

Because the receiver bases its calculation on the amount of time it takes the GNSS signal to travel from the satellite to the receiver, any errors in the satellite clocks will affect the range measurement. The satellite clock bias is the difference between the satellite clock time and the true GPS/GNSS time. Despite the fact that high quality atomic clocks (accurate to a nanosecond) are used in the GNSS satellites, the satellite clock bias is still unavoidable. Satellite clock bias can cause deviation of up to 10^{-8} seconds.

[7] The satellite orbit information is generated from the tracking data collected by the GNSS Control Segment (see Fig. 2.1 in Chap. 2). The Control Segment updates satellite positions on a regular basis, calculates their predicted paths, and uploads this information to the satellites. The uploaded information includes the *almanac* and *ephemeris* data, which contains the predicted positions of the satellites. These are downloaded when GNSS receivers track the satellite signals.

Similar to the satellite clock bias, the receiver clock bias is an offset between the receiver clock time and the true GPS/GNSS time. Because GNSS receivers are usually equipped with relatively inexpensive clocks, the receiver clock bias is much larger compared to the satellite clock bias. A time scale such as GPS time is based on highly accurate atomic clocks, while GNSS receivers contain inexpensive clocks, similar to those in some consumer devices such as digital watches. The atomic clocks are precise enough to determine time up to 11 decimal places (e.g., 10:05:46.01234567890), while a GNSS receiver's clock is less precise and may only determine time to six decimal places (e.g., 10:05:46.012345). The slight difference between the two, as a result of rounding off, is due to the poorer precision of the receiver clock.

5.2.5 Antenna Phase Center Variation

In GNSS geodesy, the measurements made by the receiver are usually referred to the distance between the electrical center of the satellite's transmitter and the electrical center of the receiver's antenna. The discrepancy between the electrical center and the physical (geometric) center is the phase center offset. The electrical center tends to vary with the direction and strength of the incoming signal. In addition, the phase center variations for the carriers (e.g., L1, L2, L5) may have different properties. For most antenna types, the antenna phase variation is usually calibrated by the manufacturers. In addition, geodetic antenna models are available from scientific and governmental organizations, such as the US NOAA's National Geodetic Survey (NGS). These models can be applied to mitigate the effect of antenna phase center variations. See further discussions in Sect. 5.3.2.3.

5.2.6 Other Factors

5.2.6.1 Hardware Biases

Both code and phase observables are biased by delays induced by the receiver and satellite hardware. The various code and phase biases are shown in the observation modeling Eqs. (5.3) and (5.4), respectively (in Sects. 5.1.1 and 5.1.2). Such biases affect the ability to resolve integer ambiguities in Precise Point Positioning (PPP), PPP-RTK, and relative carrier phase positioning when measurements from multiple GNSS systems are used. An overview of the various biases in multi-GNSS is presented in Sect. 5.3.2.5, and further details can be found in [10, 11, 21] and several other sources.

5.2.6.2 Sagnac Effect

The Sagnac effect due to the Earth's rotation during the signal propagation time from the satellite to receiver is also an important consideration when dealing with error sources in precise positioning. It is caused by the fact that a receiver on the Earth is on a fixed coordinate system that rotates with the Earth during signal propagation time. The effect implies that the onboard GNSS clock runs faster or slower relative to a clock located on, or close to, the Earth surface. The satellite's pseudorange calculation from speed of light assumes the Cartesian distance (Sect. 3.2.1 of Chap. 3) between the satellite at time of transmission and the receiver at time of reception. The satellite and receiver positions must be in the same reference frame in order to calculate the correct distance. Solving the orbital equations of satellite motion from the transmitted ephemeris data (Appendix A.2) gives the satellite's X-Y-Z position at the time of transmission. The receiver must therefore add a correction due to the rotation of ECEF frame during the signal propagation. Details of the Sagnac correction can be found in [2] and [32].

5.2.6.3 Receiver Noise

The magnitude of the receiver noise is dependent on parameters such as the signal-to-noise ratio and tracking bandwidth. A rule of thumb is that the measurement noise is approximately 1% of the signal wavelength. Therefore, the level of noise in pseudorange measurements is about 3 m (∼300 m wavelength) for C/A code and of the order of 0.3 m (∼30 m wavelength) for P code, while the level of noise in carrier phase is a few millimeters for L1 (∼19 cm wavelength) and L2 (∼24 cm wavelength). Modern receiver technology tends to bring the internal phase noise below 1 mm and to reduce the C/A code noise to the decimeter level.

5.3 Error Mitigation Methods

5.3.1 Differenced Observables

Several terms in the GNSS observation equations (5.3) and (5.4) only depend on the station or on the satellite. For example, receiver clock error is the same in all observations taken simultaneously by the same receiver. Similarly, satellite clock error is the same in all simultaneous observations from the same satellite. Such systematic errors are easily eliminated by forming mathematical differences between observations, although not in all cases for some of the terms such as atmospheric effects and orbital biases.

The typical building block for high-precision GNSS involves two stations (receivers) tracking multiple satellites (Fig. 5.1), from which three types of differ-

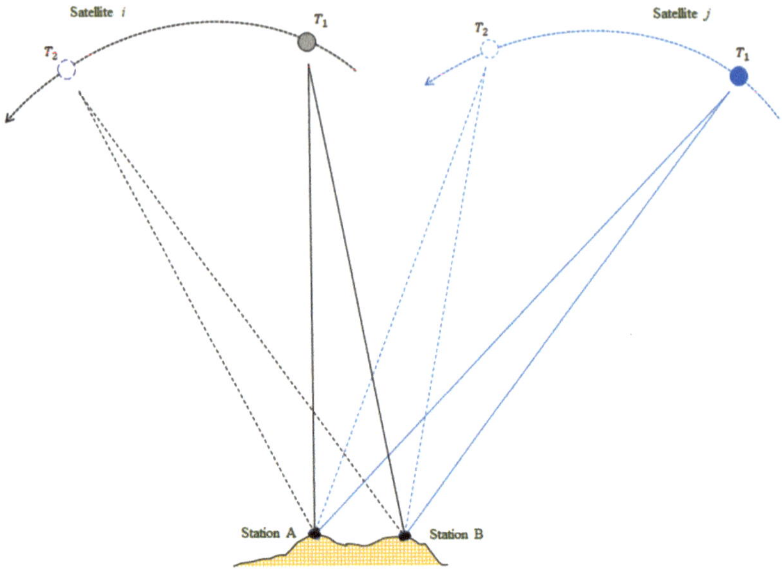

Fig. 5.1 Geometry of Satellite Observations for two stations tracking two satellites (12 single differences, 6 double differences, and 1 triple difference)

enced observables[8] can be formed: (1) *single differences* from two simultaneous observations involving either two stations (receivers) or two satellites, or two consecutive observation epochs; (2) *double differences* by taking the difference between two single differences; and (3) *triple differences* as the difference of two double differences over two epochs. The original observations (Eqs. (5.3) and (5.4)) are sometimes referred to as *zero differences* or undifferenced observations. In Fig. 5.1, there are 8 original observations, 12 single differences, 6 double differences, and 1 triple difference observable.

It is also noted here that some of the terms in the original observations, for example, γ (relativistic effects) and W (phase wind-up), can be captured by known models and are usually not estimated during GNSS processing. Such terms are omitted when forming differenced observables. Similarly, multipath is omitted when forming differences, with the assumption that its unmodeled effects contribute to the error term υ such that $\upsilon \overset{def}{=} \upsilon + M$. The antenna eccentricities $\boldsymbol{E}$ and $\boldsymbol{e}$ are constant for all types of observations and are thus included in the geocentric positions, by setting $\boldsymbol{P} \overset{def}{=} \boldsymbol{P} + \boldsymbol{E}$ and $\boldsymbol{p} \overset{def}{=} \boldsymbol{p} + \boldsymbol{e}$, when forming the differenced observation models.

---

[8] The noise of a single, double, and triple difference is a factor of $\sqrt{2}$, $\sqrt{4}$, and $\sqrt{8}$ higher compared to the undifferenced observations. All differenced observations are mathematically correlated (see, e.g., [13]).

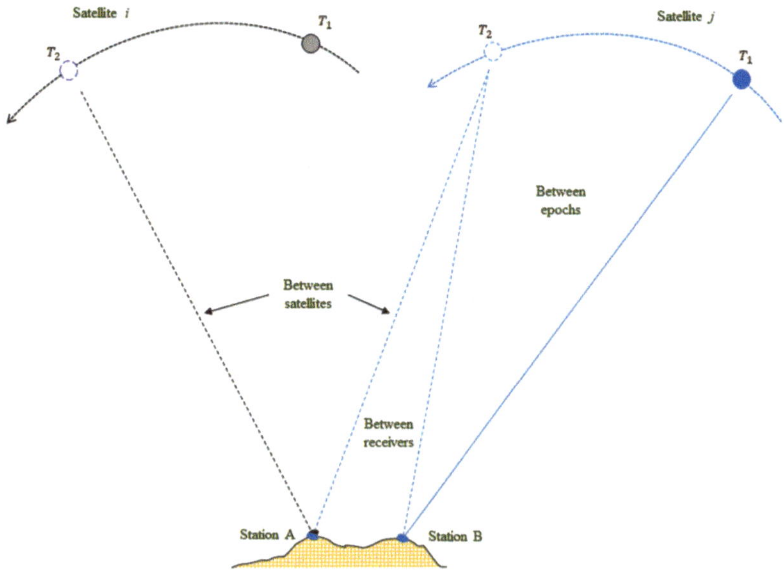

Fig. 5.2 Geometry of single differences between Satellite Observations

5.3.1.1 Single Differences

Single differences (SDs) can be formed as illustrated in Fig. 5.2, namely between-receiver differences (station differences), between-satellite differences (satellite differences), and between-epoch differences (epoch differences). Starting with the general form, single difference observation equations for two code observations (R_1, R_2) and two phase observations (Φ_1, Φ_2) are, respectively, described by the following two equations:

$$\begin{aligned} R_1 - R_2 = {} & |(\boldsymbol{P}_1 + \boldsymbol{O}_1) - (\boldsymbol{p}_1 + \boldsymbol{o}_1)| - |(\boldsymbol{P}_2 + \boldsymbol{O}_2) - (\boldsymbol{p}_2 + \boldsymbol{o}_2)| \\ & + (c\delta t_{rcv} - c\delta t^{sat})_1 - (c\delta t_{rcv} - c\delta t^{sat})_2 + cB_1 - cB_2 \\ & - cb_1 + cb_2 + d_{trop(1)} - d_{trop(2)} + d_{ion(1)} - d_{ion(2)} \\ & + d_{orb(1)} - d_{orb(2)} + v \end{aligned} \tag{5.5}$$

$$\begin{aligned} \Phi_1 - \Phi_2 = {} & |(\boldsymbol{P}_1 + \boldsymbol{O}_1) - (\boldsymbol{p}_1 + \boldsymbol{o}_1)| - |(\boldsymbol{P}_2 + \boldsymbol{O}_2) - (\boldsymbol{p}_2 + \boldsymbol{o}_2)| \\ & + (c\delta t_{rcv} - c\delta t^{sat})_1 - (c\delta t_{rcv} - c\delta t^{sat})_2 + cB_1 - cB_2 \\ & - cb_1 + cb_2 + d_{trop(1)} - d_{trop(2)} - d_{ion(1)} + d_{ion(2)} \\ & + d_{orb(1)} - d_{orb(2)} + \lambda_1(\phi_{rcv} - \phi^{sat})_1 - \lambda_1 N_1 \\ & - \lambda_2(\phi_{rcv} - \phi^{sat})_2 + \lambda_2 N_2 + v \end{aligned} \tag{5.6}$$

The two equations can be reduced to a more compact form as follows:

$$R_{12} = \rho_{12} + (c\delta t_{rcv} - c\delta t^{sat})_{12} + cB_{12} - cb_{12} + d_{trop(12)} + d_{ion(12)} + d_{orb(12)} + v \tag{5.7}$$

$$\begin{aligned}\Phi_{12} = \rho_{12} + (c\delta t_{rcv} - c\delta t^{sat})_{12} + cB_{12} - cb_{12} + d_{trop(12)} - d_{ion(12)} \\ + d_{orb(12)} + \lambda_1(\phi_{rcv} - \phi^{sat})_1 - \lambda_1 N_1 \\ - \lambda_2(\phi_{rcv} - \phi^{sat})_2 + \lambda_2 N_2 + v\end{aligned} \tag{5.8}$$

where $\rho_{12} = |(\boldsymbol{P}_1 + \boldsymbol{O}_1) - (\boldsymbol{p}_1 + \boldsymbol{o}_1)| - |(\boldsymbol{P}_2 + \boldsymbol{O}_2) - (\boldsymbol{p}_2 + \boldsymbol{o}_2)|$

Between-Receiver (Station) Differences

Given two simultaneous observations from one satellite to two receivers (at stations A and B), the code and phase between-receiver differences (i.e., station differences) are obtained as

$$\Delta R = R_{AB} = \rho_{AB} + (c\delta t_{rcv})_{AB} + cB_{AB} - cb_{AB} + d_{trop(AB)} + d_{ion(AB)} + d_{orb(AB)} + v \tag{5.9}$$

$$\Delta\Phi = \Phi_{AB} = \rho_{AB} + (c\delta t_{rcv})_{AB} + cB_{AB} + d_{trop(AB)} - d_{ion(AB)} + d_{orb(AB)} + \lambda(\phi_{rcv})_{AB} + \lambda N_{AB} + v \tag{5.10}$$

where the signal wavelengths are identical (i.e., $\lambda = \lambda_A = \lambda_B$, $\lambda N_{AB} = -\lambda N_A + \lambda N_B$); the satellite clock correction δt^{sat}, the satellite phase bias b, and the initial satellite phase reading ϕ^{sat} are the same for both observations, i.e., $(\delta t^{sat})_A = (\delta t^{sat})_B$, $b_A = b_B$, and $(\phi^{sat})_A = (\phi^{sat})_B$; $d_{trop(AB)} \approx 0$, $d_{ion(AB)} \approx 0$ and $d_{orb(AB)} \approx 0$ for a short inter-station distance. The term cB_{AB} (receiver biases)[9] is especially useful for FDMA-based observables from different receiver types. For such, it is modeled as the difference between the inter-frequency biases[10] for two receivers [33].

The presence of the satellite code bias term b_{AB} in Eq. (5.9) is an important difference between the code and phase station difference observables. Additional important points to be noted include: (a) code type and tracking mode may be different for two simultaneous observations and (b) observations by two receivers

[9] Estimating/fixing correct ambiguity depends on the bias term being introduced as additional parameter. If ignored, they are absorbed into ambiguity estimates making it difficult to fix them to correct integers. See Sect. 7.3 of Chap. 7.

[10] The GLONASS inter-frequency bias is commonly defined as the difference of bias at frequency number k with respect to the bias at frequency number 0 [29].

are usually not recorded at exactly the same epoch, but satellite clock correction and the satellite phase bias b are constant over very short time intervals (such as < 0.05 seconds).

Between-Satellite Differences

Given two simultaneous observations from two satellites i and j to one receiver, the code and phase between-satellite differences are obtained as

$$\nabla R = R^{ij} = \rho^{ij} - (c\delta t^{sat})^{ij} + cB^{ij} - cb^{ij} + d^{ij}_{trop} + d^{ij}_{ion} + d^{ij}_{orb} + v \tag{5.11}$$

$$\begin{aligned}\nabla \Phi = \Phi^{ij} &= \rho^{ij} - (c\delta t^{sat})^{ij} + cB^{ij} - cb^{ij} + d^{ij}_{trop} - d^{ij}_{ion} + d^{ij}_{orb} \\ &+\lambda^i(\phi_{rcv} - \phi^i) - \lambda^j(\phi_{rcv} - \phi^j) - \lambda^i N^i + \lambda^j N^j + v\end{aligned} \tag{5.12}$$

The receiver (station) clock synchronization terms δt^i_{rcv} and δt^j_{rcv} are the same for both observations and therefore cancel out. The initial phase reading ϕ_{rcv} is the same for all observations of a particular receiver ($\phi_{rcv} = \phi^i_{rcv} = \phi^j_{rcv}$). Further benefits are realized *if the observations are identical in frequency, code type, and tracking mode*: the receiver code bias difference B^{ij} in Eq. (5.11) is zero; the initial satellite phase terms are reduced to $-\lambda\phi^{ij}$; the ambiguity difference $-\lambda^i N^i + \lambda^j N^j$ reduces to an integer SD ambiguity λN^{ij}. The receiver phase bias difference B^{ij} in Eq. (5.12) also cancels out *if the two satellites i and j are in the same system (constellation).*[11]

Between-Epoch Differences

Given observations from one satellite to one receiver at two different epochs (T_1 and T_2), the code and phase between-epoch differences are obtained as

$$\begin{aligned}R(T_{12}) &= \rho(T_{12}) + d_{trop}(T_{12}) + d_{ion}(T_{12}) \\ &+d_{orb}(T_{12}) + c\delta t_{rcv}(T_{12}) - c\delta t^{sat}(T_{12}) + v\end{aligned} \tag{5.13}$$

$$\begin{aligned}\Phi(T_{12}) &= \rho(T_{12}) + d_{trop}(T_{12}) - d_{ion}(T_{12}) \\ &+d_{orb}(T_{12}) + c\delta t_{rcv}(T_{12}) - c\delta t^{sat}(T_{12}) + v\end{aligned} \tag{5.14}$$

Since frequency, wavelength, code type, and tracking mode are the same for both observations, most of the bias and error terms are eliminated and $d_{trop}(T_{12}) \approx 0$ and $d_{ion}(T_{12}) \approx 0$ for short epoch separations (e.g., $T_{12} \leq 30$ seconds). Equations (5.13)

[11] Applies to CDMA-based satellites.

and (5.14) contain only the variations of the station, atmospheric, and clock parameters in time.

5.3.1.2 Double Differences

Double difference (DD) observation is the difference between two single differences, for example, a code double difference is the difference between two code single differences, and a phase double difference is the difference between two phase single differences. Three types of DD observations are possible:

1. *Between-station and between-satellite double differences* involving two stations (A, B), two satellites (i, j), and one observation epoch. The DD observations are obtained from two simultaneous between-receiver (station) single differences to two satellites or from two simultaneous between-satellite single differences for two stations.
2. *Between-epoch double differences* involving two stations, one satellite and two epochs. The DD observations are obtained from two between-receiver (station) differences or two between-epoch differences.
3. *Between-epoch double differences* involving one station, two satellites, and two epochs. The DD observations are obtained from two between-satellite differences or two between-epoch differences.

The general code and phase DD observations are written as

$$\begin{aligned} R_{12} - R_{34} = {} & |(\boldsymbol{P}_1 + \boldsymbol{O}_1) - (\boldsymbol{p}_1 + \boldsymbol{o}_1)| - |(\boldsymbol{P}_2 + \boldsymbol{O}_2) - (\boldsymbol{p}_2 + \boldsymbol{o}_2)| \\ & - |(\boldsymbol{P}_3 + \boldsymbol{O}_3) - (\boldsymbol{p}_3 + \boldsymbol{o}_3)| + |(\boldsymbol{P}_4 + \boldsymbol{O}_4) - (\boldsymbol{p}_4 + \boldsymbol{o}_4)| \\ & + d_{trop(1)} - d_{trop(2)} - d_{trop(3)} + d_{trop(4)} + d_{ion(1)} - d_{ion(2)} \\ & - d_{ion(3)} + d_{ion(4)} + d_{orb(1)} - d_{orb(2)} - d_{orb(3)} + d_{orb(4)} \\ & + (c\delta t_{rcv} - c\delta t^{sat})_1 - (c\delta t_{rcv} - c\delta t^{sat})_2 \\ & - (c\delta t_{rcv} - c\delta t^{sat})_3 + (c\delta t_{rcv} - c\delta t^{sat})_4 \\ & + cB_1 - cB_2 - cB_3 + cB_4 - cb_1 + cb_2 + cb_3 - cb_4 + v \end{aligned} \tag{5.15}$$

$$\begin{aligned}\Phi_{12} - \Phi_{34} = & |(\boldsymbol{P}_1 + \boldsymbol{O}_1) - (\boldsymbol{p}_1 + \boldsymbol{o}_1)| - |(\boldsymbol{P}_2 + \boldsymbol{O}_2) - (\boldsymbol{p}_2 + \boldsymbol{o}_2)| \\ & -|(\boldsymbol{P}_3 + \boldsymbol{O}_3) - (\boldsymbol{p}_3 + \boldsymbol{o}_3)| + |(\boldsymbol{P}_4 + \boldsymbol{O}_4) - (\boldsymbol{p}_4 + \boldsymbol{o}_4)| \\ & +d_{trop(1)} - d_{trop(2)} - d_{trop(3)} + d_{trop(4)} - d_{ion(1)} + d_{ion(2)} \\ & +d_{ion(3)} - d_{ion(4)} + d_{orb(1)} - d_{orb(2)} - d_{orb(3)} + d_{orb(4)} \\ & +(c\delta t_{rcv} - c\delta t^{sat})_1 - (c\delta t_{rcv} - c\delta t^{sat})_2 \\ & -(c\delta t_{rcv} - c\delta t^{sat})_3 + (c\delta t_{rcv} - c\delta t^{sat})_4 \\ & +cB_1 - cB_2 - cB_3 + cB_4 - cb_1 + cb_2 + cb_3 - cb_4 \\ & +\lambda_1(\phi_{rcv} - \phi^{sat})_1 - \lambda_1 N_1 - \lambda_2(\phi_{rcv} - \phi^{sat})_2 + \lambda_2 N_2 \\ & -\lambda_3(\phi_{rcv} - \phi^{sat})_3 + \lambda_3 N_3 + \lambda_4(\phi_{rcv} - \phi^{sat})_4 - \lambda_4 N_4 + v \end{aligned} \tag{5.16}$$

Between-Station and Between-Satellite Double Differences

In this type of double difference, the receiver and clock offsets δt_{rcv} and δt^{sat}, the satellite phase biases b, and the initial phase readings ϕ^{sat} (in this case ϕ^i and ϕ^j) are eliminated. Subsequently, the reduced forms of code and phase DD observations are as follows:

$$\Delta\nabla R = R^{ij}_{AB} = \rho^{ij}_{AB} + d^{ij}_{trop(AB)} + d^{ij}_{ion(AB)} + cB^{ij}_{AB} + cb^{ij}_{AB} + v \tag{5.17}$$

$$\begin{aligned}\Delta\nabla\Phi = \Phi^{ij}_{AB} = & \rho^{ij}_{AB} + d^{ij}_{trop(AB)} - d^{ij}_{ion(AB)} + cB^{ij}_{AB} \\ & +\lambda^i(\phi_{rcv})^i_{AB} - \lambda^j(\phi_{rcv})^j_{AB} + \lambda^i N^i_{AB} - \lambda^j N^j_{AB} + v\end{aligned} \tag{5.18}$$

The benefits of this type of DD observations include: (i) eliminating receiver and satellite clock biases and (ii) reducing the number of parameters to be estimated. The term $\lambda^i N^i_{AB} - \lambda^j N^j_{AB}$ in Eq. (5.18) gives an integer DD ambiguity *if the carrier wavelength* λ *is the same for all observations* (thus, a challenge for multi-GNSS case such as GPS+GLONASS). The initial phase terms ϕ_{rcv} also cancel out, and the receiver phase biases B are eliminated *if frequency and constellations are the same* (and are CDMA-based). In Eq. (5.17), the biases b are eliminated *if the same code types are observed by both receivers*, and receiver biases B are eliminated *if both receivers use the same code type and tracking mode for the code measurements*. For very short baselines, the terms $d^{ij}_{trop(AB)}$ and $d^{ij}_{ion(AB)}$ will be approaching zero.

Between-Epoch Double Differences

The code and phase observation equations for between-epoch double differences involving two stations (A, B), one satellite, and two epochs (T_1, T_2) can be written as follows:

$$R_{AB}(T_{12}) = \rho_{AB}(T_{12}) + d_{trop(AB)}(T_{12}) + d_{ion(AB)}(T_{12}) + c\delta t_{rcv(AB)}(T_{12}) + v \tag{5.19}$$

$$\Phi_{AB}(T_{12}) = \rho_{AB}(T_{12}) + d_{trop(AB)}(T_{12}) - d_{ion(AB)}(T_{12}) + c\delta t_{rcv(AB)}(T_{12}) + v \tag{5.20}$$

The code and phase observation equations for between-epoch double differences involving one station, two satellites (i, j), and two epochs (T_1, T_2) can be written as follows:

$$R^{ij}(T_{12}) = \rho^{ij}(T_{12}) + d^{ij}_{trop)}(T_{12}) + d^{ij}_{ion}(T_{12}) + c\delta t^{ij}_{rcv}(T_{12}) + v \tag{5.21}$$

$$\Phi^{ij}(T_{12}) = \rho^{ij}(T_{12}) + d^{ij}_{trop)}(T_{12}) - d^{ij}_{ion}(T_{12}) + c\delta t^{ij}_{rcv}(T_{12}) + v \tag{5.22}$$

In both types of double differences, the following assumptions are made: (i) the difference between the two observation epochs T_1 and T_2 is small; and (ii) the frequency, wavelength, code type, and tracking mode are the same for both epochs. With these assumptions, receiver and satellite biases, phase ambiguities, and initial phase readings are eliminated.

5.3.1.3 Triple Differences

Triple differencing is useful for preliminary baseline computations and pre-processing purposes such as cycle slip detection and repair (Table 5.1). It is obtained by differencing two double differences between epochs. A triple difference observable involves exactly two stations (A, B), two satellites (i, j), and two observation epochs (T_1, T_2). Thus the code and phase observation equations are, respectively, written as follows:

$$R^{ij}_{AB}(T_{12}) = \rho^{ij}_{AB}(T_{12}) + d^{ij}_{trop(AB)}(T_{12}) + d^{ij}_{ion(AB)}(T_{12}) + v \tag{5.23}$$

$$\Phi^{ij}_{AB}(T_{12}) = \rho^{ij}_{AB}(T_{12}) + d^{ij}_{trop(AB)}(T_{12}) - d^{ij}_{ion(AB)}(T_{12}) + v \tag{5.24}$$

Table 5.1 Different ways to mitigate systematic errors

Observable	Overall impact on eliminating/reducing errors
Between-satellite single difference (SD) (see Fig. 5.2)	Eliminates receiver clock bias. Ambiguity term is float due to phase bias.[a]
Between-receiver single difference (SD) (see Fig. 5.2)	Eliminates satellite clock bias. Greatly reduces ephemeris, ionospheric, and tropospheric biases (if baseline is short).
Between-epoch single difference (SD) (see Fig. 5.2)	Eliminates carrier phase ambiguity term. Greatly reduces ephemeris, ionospheric, and tropospheric biases (if epoch separation is short)
Receiver-satellite double difference (DD)	Eliminates receiver and satellite clock biases. Significantly reduces effect of ephemeris, ionospheric, and tropospheric biases (if baseline is short).
Triple difference (TD)	Eliminates carrier phase ambiguity term. Suited for cycle slip detection/repair.

[a]The ambiguity term is still float since it contains initial phase bias but can be integer if phase bias is zero or if the size of bias is known

The clock error terms, receiver and satellite biases,[12] initial phase readings, and ambiguities are all eliminated. The terms $d^{ij}_{trop(AB)}(T_{12})$ and $d^{ij}_{ion(AB)}(T_{12})$ will be approaching zero for short inter-station distances and short epoch intervals (e.g., $T_{12} = 1$ second).

5.3.2 *Error Modeling*

5.3.2.1 Clock Corrections Modeling

The satellite and receiver clock offsets (δt^{sat}, δt_{rcv}) presented earlier in the code and phase observation equations (5.3) and (5.4) can be handled in different ways. In most instances, data differencing methods (Sect. 5.3.1) can eliminate or reduce their impacts on parameter estimation. However, there are also empirical models that can be applied during data processing to correct for some of these offsets as applicable, for example, in some instances of single difference solutions. While receiver clock offsets are typically estimated as nuisance parameters along with the position coordinates, the satellite clock offsets can be estimated from models. Estimates of receiver clock offsets (and drifts) are available within the receiver; hence, manufacturers typically adjust estimate of receiver time, e.g., by resetting

[12] Code and phase biases (B, b) caused by receiver and satellite hardware delays of signals (see Eqs. (5.3) and (5.4)).

the clock when the offset exceeds a threshold ("millisecond jumps," see [23] for details).

One way to account for the satellite clock offsets is to use the broadcast clock error model defined by polynomial coefficients (e.g., as generated by the GPS Master Control Station). It should, however, be noted that even with the best efforts in monitoring the behavior of each satellite clock, their behavior cannot be precisely predicted. Consequently, there is a residual error after applying the broadcast clock error model. However, as shown in Sect. 5.3.1, the satellite clock bias can be eliminated by differencing the measurements obtained from two receivers since the satellite clock offset is the same for two receivers observing the same satellite, at the same time.

The satellite clock offset δt^{sat} can be split into two terms [30]:

$$\delta t^{sat} = \widetilde{\delta t}^{sat} + \Delta_{rel} \tag{5.25}$$

The first term can be calculated from the polynomial coefficients in broadcast navigation messages, or from precise IGS products,[13] or similar products by other providers of such information. The second term is a small relativistic correction caused by the satellite's orbital eccentricity.

Broadcast navigation message (i.e., broadcast ephemeris) usually contains satellite clock information in the form of polynomial coefficients (a_0, a_1, a_2) in a given reference epoch t_0, to compute the satellite clock offset. With these information, the first term in Eq. (5.25) is computed as

$$\widetilde{\delta t}^{sat} = a_0 + a_1(t - t_0) + a_2(t - t_0)^2 \tag{5.26}$$

where the satellite clock offset (a_0), clock drift (a_1), and clock drift rate (a_2) are the broadcast ephemeris polynomials (typically renewed every two hours and must not be used after about four hours, because extrapolation error grows exponentially beyond the prescribed validity period). Equation (5.26) applies for GPS, Galileo, and BeiDou satellites, for which the broadcast message is valid for two hours. For GLONASS satellites, however, the broadcast message is updated every half-hour, and therefore, only a first-order polynomial (with the first two terms) is considered, i.e., the clock offset $(a_0 = -\tau_n)$ and the relative frequency offset $(a_1 = \gamma_n)$.[14]

The second term (Δ_{rel}) in Eq. (5.25) is a periodic component of the *relativistic clock correction*,[15] which is typically applied by receiver firmware [1]. It is added to a constant component that depends only on the nominal value of the semi-major

[13] The precise GPS and GLONASS satellite clocks provided by IGS are accurate to the order of 0.1 ns or better (1 ns of error $\approx$30 cm in range) [30].

[14] GLONASS satellites transmit Δ_{rel} within the satellite clock corrections $\widetilde{\delta t}^{sat}$ [30].

[15] The rate of advance of two identical clocks, one on the satellite and the other on ground, will differ due to the difference in gravitational potential (general relativity) and the relative speed between them (special relativity).

axis of the satellite orbit [30]. The constant component is adjusted by modifying (in the factory) the clock oscillating frequency of the satellite such that, for example:

$$\frac{f_0' - f_0}{f_0} \simeq 4.464 \cdot 10^{-10} \tag{5.27}$$

where f_0' is the frequency emitted by the satellite and f_0 is the one received on ground. For $f_0 = 10.23\,\text{MHz}$, $\Delta f_0 = 4.464 \cdot 10^{-10} \cdot f_0 = 4.57 \cdot 10^{-3}\,\text{Hz}$,[16] thus the satellite must use $f_0' = 10.22999999543\,\text{MHz}$.

5.3.2.2 Atmospheric Effects Modeling

In double and triple difference observations (Sect. 5.3.1.2 and 5.3.1.3), the tropospheric and ionospheric effects will approach zero for short inter-station (baseline) distances. However, these atmospheric effects become problematic for longer baselines and/or situations where two stations have steep elevation differences between them (even if their baseline is short or medium range, e.g., 1 km or longer). Figure 5.3 illustrates the two scenarios.

Considering a simple case of identical signals traveling from the same satellite to two ground stations A and B (as shown in Fig. 5.3):

1. In long baselines (inter-station distances), the two signals travel different distances through the atmosphere and encounter different atmospheric conditions (both in the ionosphere and troposphere), even if the elevation difference between stations A and B is small.
2. In short baselines (inter-station distances), the two signals travel nearly similar (or equal) distances through the ionosphere and encounter similar ionospheric effect (delay) but encounter different troposphere layers between the two stations A and B if the elevation difference between them is steep. Otherwise, the two signals travel nearly similar (or equal) distances through the atmosphere and encounter similar atmospheric conditions (in the ionosphere and troposphere).

Because of the above realities, it becomes necessary to consider (and apply) empirical models for atmospheric effects on GNSS signals in order to remove, or reduce, their impact on parameter estimation, and more so when dealing with long baselines or locations with steep elevations.

Using dual (or triple)-frequency GNSS measurements becomes advantageous in handling some of these effects. For example, dual-frequency receivers can eliminate first-order ionospheric effects through a linear combination of code or phase measurements (see Appendix B).

---

[16] The clock on the satellite appears to run faster by $\approx 38\,\mu\text{s/day}$ than on ground (since $\Delta f/f = \Delta T/T$). This effect is corrected (in the factory) by decreasing the oscillating frequency of the satellite by the amount $4.57 \cdot 10^{-3}$ Hz.

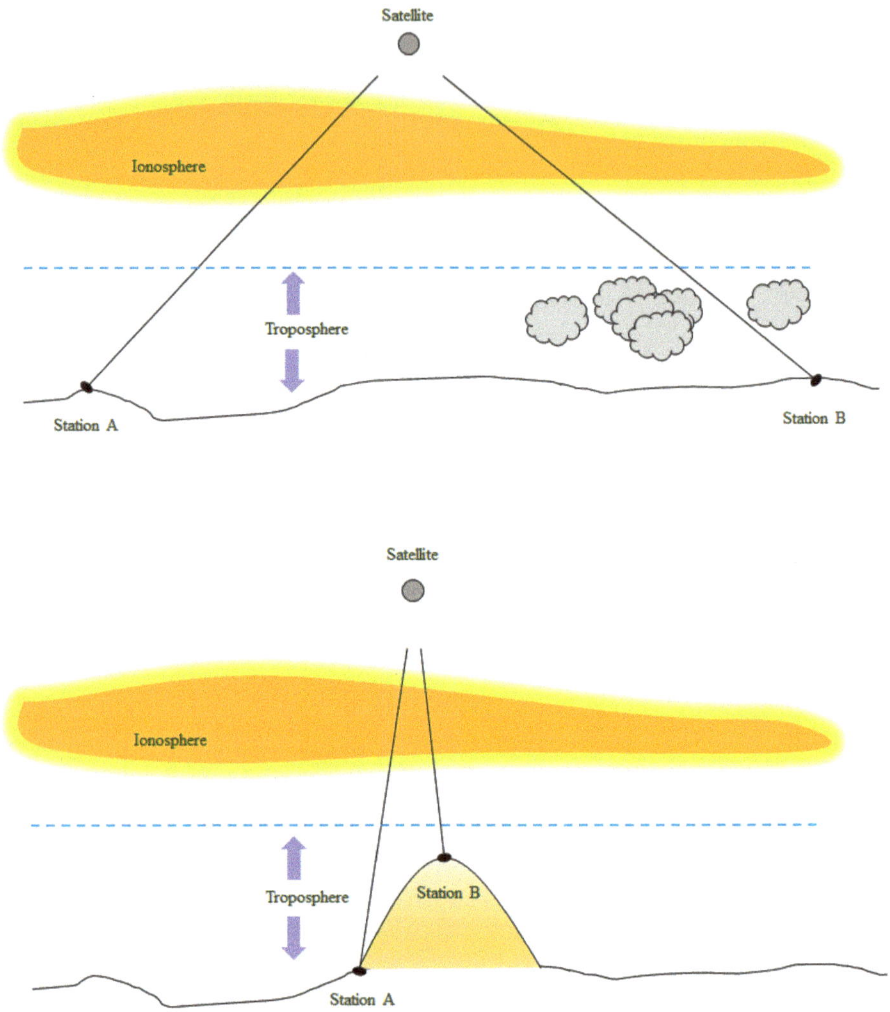

Fig. 5.3 Atmospheric effects on GNSS signals

Correction for Ionospheric Delay

Ionospheric delays are highly correlated over distances of up to a few tens of kilometers; therefore, the impact of ionospheric delay can be significantly reduced by forming a difference between measurements made by two receivers, on a short baseline, to the same satellite. This method is already implied in Sects. 5.3.1.2 and 5.3.1.3 above.

The ionosphere is dispersive[17] for GNSS signals (refraction of the signals depends on (varies as) the squared inverse of their frequencies) such that the signal delay for a higher frequency carrier is less than that of a lower frequency carrier. This dependence on the signal frequency leads to the advantage of dual- and triple-frequency receivers over single frequency ones. For example, through the several possible two-frequency combinations (e.g., L1/L2, L1/L5, or L2/L5), it allows to remove the effects by up to more than 99.9% in the ionosphere-free linear combination.

In other cases, such as methods employing single frequency measurements, or where data differencing is not effective, ionospheric prediction models can be applied to estimate and remove the effects of *group delay* and *phase delay*[18] (i.e., the term d_{ion} in Eq. (5.3) and $-d_{ion}$ in Eq. (5.4), respectively). The models take into account both first-order ($I_{(1)f}$) and second-order ($I_{(2)f}$) ionospheric effects such that $d_{ion} = I_{(1)f} + I_{(2)f}$ for a GNSS signal of carrier frequency f [13, 28, 30]. The first-order ionospheric effects are often sufficient for estimating the most significant part of the delays, as the remaining contributions often only amount to a few mm [17].

At the first order, the relationship between the ionospheric signal delay and the carrier frequency f (in Hz) can be expressed as [9, 15]

$$I_{(1)f,R} = \frac{40.3 \cdot STEC}{f^2} \quad I_{(1)f,\Phi} = -\frac{40.3 \cdot STEC}{f^2} \tag{5.28}$$

where the first and second equations are the delays for code and phase observations, respectively. $STEC$ is the slant total electron content (TEC),[19] which represents the TEC along the signal propagation path (Fig. 5.4). The two terms $I_{(1)f,R}$ and $I_{(1)f,\Phi}$ (ionospheric delays for code and phase) have different signs. This simply means the code is delayed, while the carrier phase is advanced (i.e., speeds up in the ionosphere).

The code and carrier terms for second-order ionospheric effects are similarly related according to the equation [12]:

$$I_{(2)f,R} = -2I_{(2)f,\Phi} \tag{5.29}$$

[17] The ionosphere is dispersive for GNSS signals, which means that the signal delays differ depending on the carrier frequency employed.

[18] GNSS carrier waves propagate with the phase velocity, whereas code measurements are considered to propagate with group velocity (i.e., pseudoranges obtained from the codes modulated in the carriers); the carrier waves speed up in the ionosphere, affected by what is known as the phase delay, while code measurements appear to be delayed or slowed by what is known as the group delay.

[19] TEC is the number of free electrons expressed in TEC units (TECUs), where 1 TECU = 10^{16} electrons per m^2.

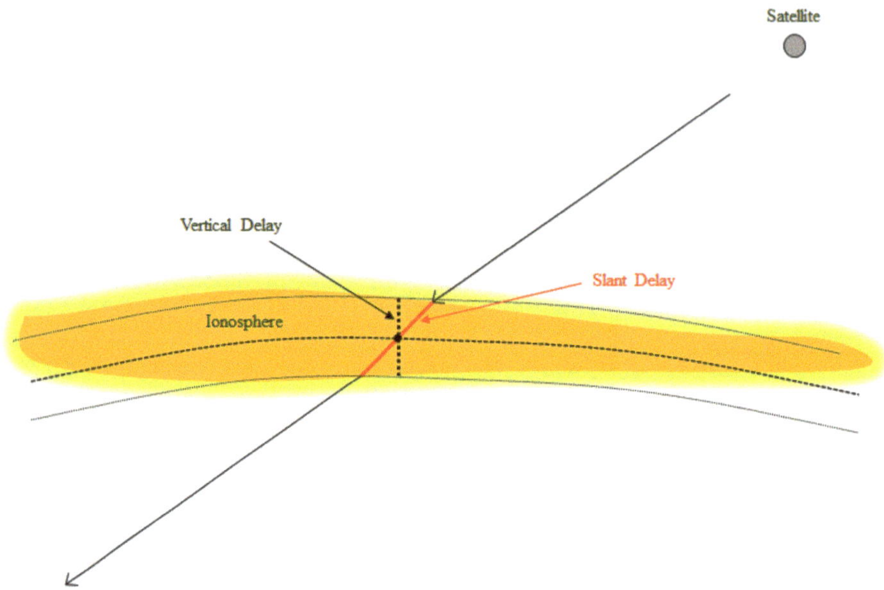

Fig. 5.4 Illustration of vertical and slant ionospheric delay

The second-order ionospheric terms mainly affect the satellite clock estimate (at centimeter level) and orbits (a few mm), but the impact on receiver position is typically less than 1 mm in global geodetic computations.

In Eqs. (5.28) and (5.29), the TEC depends on geographic location, time of day, and intensity of solar activity.

Correction for Tropospheric Delay

The main feature of the troposphere is that it is a non-dispersive medium with respect to GNSS signals; thus, unlike the ionospheric delay, the tropospheric delay is not frequency dependent. The code and carrier measurements are affected by the same delay. The consequence of the non-frequency dependence of the tropospheric delay is that it cannot be eliminated through linear combinations of dual- or multiple frequency observations (as is the case with the ionosphere). Several "standard" troposphere models can be used to estimate the magnitude of the tropospheric delay (e.g., Saastamoinen model, Hopfield model, Black model, and others, see, for example, [13, 17]). These models rely on the fact that tropospheric delay depends on the temperature, pressure, and humidity, as well as the transmitter and receiver antenna locations.

The modeling of the troposphere can be separated into a dry and a wet part,[20] where the dry part constitutes the hydrostatic atmosphere (dry gases, mainly N_2 and O_2) and the wet part constitutes water vapor [13]. Thus, the signal refraction due to the troposphere is separated into the dry component and the wet component. The dry component varies with the local temperature and atmospheric pressure in a very predictable manner (variation of less than 1% over a few hours). The wet component depends on the weather conditions, varies faster than the dry component, and is difficult to predict or model. Fortunately, most of the tropospheric delay (about 90%) comes from the predictable dry component [17]. As a result of this, the standard models can account for about 90% of the total delay (the dry component). It is also common, in high-accuracy positioning, to estimate the wet delay as an additional unknown together with the coordinates, while the dry part is corrected by a deterministic model [17].

As explained earlier using Fig. 5.3, tropospheric delay can be reduced or eliminated by forming a difference between the measurements made by two receivers on short baselines to the same satellite (Sects. 5.3.1.2 and 5.3.1.3). In precise geodetic computations, the residual tropospheric delays in the double-differenced observables (Sect. 5.3.1.2) may be treated as additional unknown parameters in the baseline estimation procedure.

5.3.2.3 Antenna Phase Center Modeling

As previously mentioned in Sect. 5.2.5, the distance traveled by the radio signal from a satellite to a receiver is the basis for GNSS positioning. Theoretically, that distance should be from the geometrical center of the satellite to that of the receiving antenna. However, in reality it is defined by the distance traveled from the electrical phase center of the satellite and to that of the receiving antenna. This is illustrated in Fig. 5.5. Although slightly exaggerated, the figure shows that the electrical phase center is not necessarily coincident with the physical geometrical center.

The phase center of the satellite antenna is the apparent electrical location of the signal leaving the satellite transmitter (the apparent source of radiation), whereas the phase center of the receiving antenna is the apparent electrical location at which the incoming radio signal is received at the antenna. In both cases, the location of the phase center is not necessarily the geometrical center or center of mass of the antenna, although the two would be coincident with each other in a perfect situation. Furthermore, the phase center is not constant but varies with the frequency, direction (azimuth and elevation), and intensity of the satellite signal.

In an ideal case, the phase center would have a spherical equiphase contour, but in reality the equiphase contour is irregular due to each segment (or direction) having its own apparent radiation source or origin (see, e.g., [13, 30]). Each radiating

[20] The water vapor part only covers the lowest layer of the troposphere below 13 km above the surface of the Earth, while the dry part extends to about 45 km above the surface of the Earth [9].

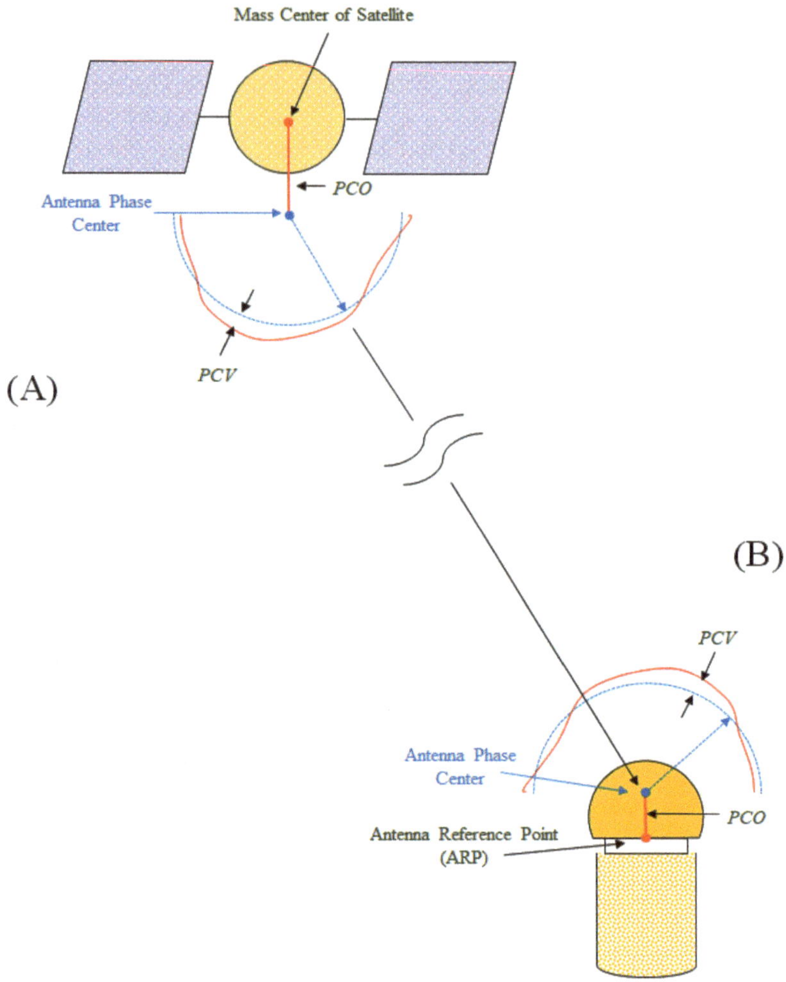

Fig. 5.5 Satellite (**a**) and receiver (**b**) antenna phase center model

or incoming satellite signal has its own electrical phase center. Therefore, a mean position of the electrical phase center is determined for the purpose of calibrating an offset.[21]

[21] The electrical phase center of the antenna is the point to which all measurements derived from GNSS signals are referred. This point cannot be accessed (e.g., by physical measurement). Therefore, a geometrical point on the antenna denoted as antenna reference point (ARP) is introduced for the purpose of defining a phase center offset (PCO)—the difference between ARP and the mean electrical phase center [13].

The phase center offset (PCO) is the difference between the ARP and the mean electrical antenna phase center (APC) and is typically represented by three-dimensional coordinates of the APC with reference to the ARP. Due to being frequency dependent, the antenna PCO is defined for each carrier frequency. The phase center variations (PCV) are the deviations that occur as a result of comparing the electrical phase center of an individual measurement (signal) with the mean electrical phase center (Fig. 5.5). The PCVs are azimuth- and elevation-dependent and thus define the antenna phase pattern. In the GNSS observation modeling, the total antenna phase center correction for an individual phase measurement (signal) is the PCO plus the azimuth- and elevation-dependent PCV [13].

Antenna phase center corrections for different GNSS antenna models are established through antenna calibration models that are routinely compiled by different scientific bodies and organizations such as the IGS and NGS. Equipment manufacturers also typically provide the technical information relating to the phase center offsets. Such information includes, for example, the North East Up offsets of the mean electrical phase center location and PCV as a function of the azimuth and elevation angle. Since circa 2006 (when ITRF2005 was adopted), there have existed two types of antenna correction models, relative and absolute APC corrections.

Finally, the satellite antenna phase center modeling is carried out a little differently than the receiving antenna phase center modeling. The satellite's broadcast ephemeris (i.e., broadcast orbits) is referred to the satellite's APC in an ECEF reference frame, and there is no additional correction needed when using the broadcast navigation message [30]. However, precise orbits and clocks are referred to the satellite's center of mass (Fig. 5.5) and thus are necessary to incorporate PCO vector when using these products.

5.3.2.4 Carrier Phase Wind-up Effect

As previously mentioned in Sect. 5.3.1, the phase wind-up term (i.e., λW in Eq. (5.4)) is usually omitted for differenced observables from short baselines.[22] However, the residual errors from this term can be on the order of a few millimeters in double-differenced observations from baselines of several hundreds of kilometers, and up to a quarter of a wavelength from longer baselines of thousands of kilometers (see, e.g., [32, pp. 569–570]). The correction is typically included in the modeling of undifferenced observations such as in PPP methods for high-accuracy point positioning.

The effect is significant for undifferenced PPP when fixing IGS satellite clocks, since it can reach up to half a wavelength [16, 34]. Most IGS Analysis Centers (i.e., IGS orbit/clock combined products) apply the correction, as ignoring it will result in position and clock errors at the decimeter level [16]. However, for receiving antenna

[22] GNSS baseline lengths of tens of kilometers or less.

rotations, such as during kinematic positioning, the phase wind-up is fully absorbed into station clock solutions or eliminated by double differencing [16].

The "phase wind-up" [23] correction (in cycles) is formulated from the dot $(\cdot)$ and vector $(\times)$ products using (see [34] and [16])

$$W = sign(\xi)cos^{-1}\left(\frac{\mathbf{D}' \cdot \mathbf{D}}{||\mathbf{D}'||\,||\mathbf{D}||}\right) \quad (5.30)$$

where $\xi = \boldsymbol{k} \cdot (\mathbf{D}' \times \mathbf{D})$, $\boldsymbol{k}$ is the satellite to receiver unit vector (pointing from the satellite transmitter to the receiver), and $\mathbf{D}'$, $\mathbf{D}$ are the effective dipole vectors of the satellite and receiver computed from the current satellite body coordinate unit vectors $(\boldsymbol{x}', \boldsymbol{y}', \boldsymbol{z}')$ and the local receiver unit vectors (i.e., north, east, up) denoted by $(\boldsymbol{x}, \boldsymbol{y}, \boldsymbol{z})$, such that:

$$\begin{aligned} \mathbf{D}' &= \boldsymbol{x}' - \boldsymbol{k}(\boldsymbol{k} \cdot \boldsymbol{x}') - \boldsymbol{k} \times \boldsymbol{y}' \\ \mathbf{D} &= \boldsymbol{x} - \boldsymbol{k}(\boldsymbol{k} \cdot \boldsymbol{x}) + \boldsymbol{k} \times \boldsymbol{y} \end{aligned} \quad (5.31)$$

A unit vector, for example $\boldsymbol{k}$, the unit vector pointing from the satellite's center of mass to the receiver, can be computed as follows:

$$\boldsymbol{k} = \frac{\mathbf{r}^{SCoM} - \mathbf{r}_{rcv}}{||\mathbf{r}^{SCoM} - \mathbf{r}_{rcv}||} \quad (5.32)$$

where $\mathbf{r}^{SCoM}$ are the satellite's center of mass (CoM) coordinates and $\mathbf{r}_{rcv}$ are the receiver's coordinates, both in the ECEF reference frame. Thus, the other unit vectors are computed in a similar manner.[24]

5.3.2.5 Code and Phase Biases

The code and phase bias terms have been introduced and discussed in Eqs. (5.3) and (5.4) as well as in the various differenced observables in Sect. 5.3.1. The biases are highly correlated with other terms such as clock errors and therefore hard to estimate in undifferenced form. However, it is possible to estimate differences between biases from code and phase observations, in various ways for various applications.

[23] GNSS "satellites transmit right circularly polarized (RCP) radio waves and therefore, the observed carrier phase depends on the mutual orientation of the satellite and receiver antennas. A rotation of either receiver or satellite antenna around its bore (vertical) axis will change the carrier-phase"[16]. The change in phase can be up to one cycle (one wavelength), which corresponds to one complete revolution of the antenna. Mostly, the receiving antenna is fixed (unless mobile), and satellite antennas undergo slow rotations as their solar panels are being oriented toward the Sun, and thus the station-satellite geometry changes causing the phase wind-up.

[24] A unit vector pointing from satellite's CoM to the geocenter is $\mathbf{r}^{SCoM}/||\mathbf{r}^{SCoM}||$.

Biases differ between different signals and between different carrier waves. It is assumed that receiver hardware biases are same for satellites belonging to same constellation and broadcasting the same signal [10]. This assumption holds true for CDMA-based constellations but does not apply in GLONASS biases for the FDMA-based satellites.

The following options are generally available for dealing with the code and phase biases in precise positioning [11]:

1. **Eliminating the satellite bias** through single differenced observables where possible, for instance, the satellite bias cancelling out in between-receivers differencing using identical signals and same wavelength (see explanations for Eqs. (5.9) and (5.10) in Sect. 5.3.1)
2. **Eliminating the receiver bias** through single differenced observables where possible, for instance, the receiver bias cancelling out in between-satellite differencing if both satellites belong to the same constellation (see explanations for Eqs. (5.11) and (5.12) in Sect. 5.3.1)
3. **Estimating the bias** as one of the unknown parameters during the data processing or positioning process (see, e.g., [33])
4. **Correcting for the bias** by using estimates available from other sources (e.g., from data service providers)
5. **Calibrating the bias** by using pre-estimated bias (by other sources) to find estimates of more stable bias

The types of code and phase biases include:

(a) Phase Intersystem Bias (ISB) in PPP and relative positioning when satellites from different constellations are used in position processing. These, in addition to time scale and reference frame differences, should be taken into account when processing multi-constellation data.
(b) Phase Inter-Frequency Bias (IFB) when FDMA-based GLONASS data are used in processing. See details in [29] and [33].
(c) Differential Code Bias (DCB) between two signals transmitted by a single satellite consists of delays induced in the receiver hardware at reception time and in satellite hardware at transmission time. They occur from the use of different carrier frequencies but also between different signal types using same carrier frequency (e.g., L1C and L1P of GPS).
(d) Code ISB and GLONASS Code IFB both which also need to be taken into account in PPP and relative positioning (see e.g., [11]).

5.3.2.6 Earth Deformation Effects

The preceding Sects. 5.3.2.1 through 5.3.2.5 have covered the error sources impacting on the satellite to receiver range estimation from the signal propagation. Modeling and/or removing such errors translate into a more precise estimation of receiver position at the measurement epoch. However, in a global sense, an Earth-based station or geodetic monument upon which a GNSS antenna is fixed/installed is part of the larger Earth surface that undergoes periodic movements (semi-diurnal, diurnal, and so forth).

Most of the periodic Earth movements are nearly the same over large areas of the Earth surface. Such periodic movements nearly cancel and need not be considered in differential positioning over short baselines (<100 km). However, for precise station coordinate solutions consistent with the ITRF conventions in relative positioning over long baselines (>500 km) and PPP (undifferenced solutions), the periodic Earth movements need to be modeled as recommended in the IERS conventions. This is accomplished by adding the site displacement correction terms to the ITRF coordinates, according to the following equation:

$$\mathbf{r}_P = \mathbf{r}_{P_0} + \Delta\mathbf{r}_{sol} + \Delta\mathbf{r}_{ocn} + \Delta\mathbf{r}_{pol} \tag{5.33}$$

where $\mathbf{r}_P$ is the station (monument) position[25] in an ECEF reference frame and $\mathbf{r}_{P_0}$ are its corresponding coordinates that are free of the periodic movements ($\Delta\mathbf{r}_{sol}$, $\Delta\mathbf{r}_{ocn}$, and $\Delta\mathbf{r}_{pol}$). The displacements $\Delta\mathbf{r}_{sol}$, $\Delta\mathbf{r}_{ocn}$, and $\Delta\mathbf{r}_{pol}$ are described by geophysical models or gridded convolution results derived from geophysical models of Earth deformations according to established IERS Conventions (see, e.g., [22]). Here are brief summaries.

Solid Earth Tides

Solid Earth tides are the horizontal and vertical displacements of the solid Earth surface due to the gravitational forces of the Sun and the Moon on the Earth's crust. They are expressed by spherical harmonics expansion of degree and order (n, m) characterized by the Love number h_{nm} and the Shida number l_{nm}, whose effective values weakly depend on site latitude and tidal frequency [16]. In a simplified model (where a position precision to a few millimeters is desired), only the site displacement vector due to the degree 2 tide is necessary. That site displacement vector in Cartesian coordinates is given by the IERS Convention's degree 2 tides displacement model—in-phase corrections (see [22, p. 103]), according to the following expression:

$$\Delta\mathbf{r}_{sol} = \sum_{j=2}^{3} \frac{GM_j R_e^4}{GM_\oplus R_j^3} \left\{ h_2\hat{\mathbf{r}}\left(\frac{3(\hat{R}_j \cdot \hat{\mathbf{r}})^2 - 1}{2}\right) + 3l_2(\hat{R}_j \cdot \hat{\mathbf{r}})\left[\hat{R}_j - (\hat{R}_j \cdot \hat{\mathbf{r}})\hat{\mathbf{r}}\right]\right\} \tag{5.34}$$

[25] The station position vector $\mathbf{r}_P$ is computed by subtracting the ARP offset vector ($\boldsymbol{\Delta}_{ARP}$), defining ARP positioning relative to the station point P (monument marker), and the APC offset vector ($\boldsymbol{\Delta}_{APC} \equiv PCO$ in Fig. 5.5), defining the APC position relative to the ARP, from the receiver's APC position $\mathbf{r}$. Thus $\mathbf{r}_P = \mathbf{r} - \boldsymbol{\Delta}_{ARP} - \boldsymbol{\Delta}_{APC}$. It is also noted that if ARP is the same as the monument marker position, then $\mathbf{r}_P = \mathbf{r} - \boldsymbol{\Delta}_{APC}$.

where:

GM_j	is the gravitational parameter for the Moon ($j = 2$) or the Sun ($j = 3$).
$GM_{\oplus}$	is the gravitational parameter for the Earth.
$\hat{R}_j, R_j$	are the unit vector from the geocenter to the Moon or the Sun and the magnitude of that vector.
R_e	is the Earth's equatorial radius.
$\hat{\mathbf{r}}, r$	are the unit vector from the geocenter to the site (station) and the magnitude of that vector.
h_2	is the nominal degree 2 Love number.
l_2	is the nominal degree 2 Shida number.

The complete model implementing up to degree 3 tides (see [30, pp. 134–136] and [22, pp. 103–107]) should be taken into account if further refinement in tidal displacement accuracy, e.g., to sub-millimeter level, is desired.

In point positioning (e.g., PPP), neglecting the solid tides correction would result in systematic position errors of up to a few centimeters [30]. However, for differential positioning over short baselines (<100 km), both stations would have almost identical tidal displacements, in which case their relative positions will be largely unaffected by the solid Earth tides.

Ocean Loading

Ocean loading is a tidal effect due to elastic response of the Earth's crust to the load of (mainly diurnal and semi-diurnal) ocean tides. The resulting deformation occurs in the ocean (or sea) floor and the adjacent lands and is therefore more localized compared to the effect of solid Earth tides.

The models for ocean tidal loading are described in the IERS Convention's technical report (see, e.g., [22, pp. 108–111]). The simplified version can be summarized as follows:

$$\Delta \mathbf{r}_{ocn} = \sum_{k=1}^{11} f_k A_{ck} cos(\chi_k(t) + u_k - \phi_{ck}) \tag{5.35}$$

where:

k	the summation of k represents the 11 tidal waves M_2, S_2, N_2, K_2, K_1, O_1, P_1, Q_1, M_f, M_m, and S_{sa}.
f_k, u_k	depend on the longitude of the lunar node and for 1–3 mm precision can be set to $f_k = 1$ and $u_k = 0$ [16].
A_{ck}, ϕ_{ck}	the amplitudes A_{ck} and phases ϕ_{ck} of the tidal loading response for the site.
$\chi_k(t)$	is an astronomical argument at time $t = 0h$, corresponding to the tidal wave component k.

The 11 tidal constituents used in the summation (Eq. (5.35)) include the semi-diurnal (M_2, S_2, N_2, and K_2) and diurnal (K_1, O_1, P_1, and $Q1$) constituents that have periods of a day or less, and long period constituents. The long period constituents include the Lunar fortnightly (M_f) and monthly (M_m) and the Solar semi-annual (S_{sa}) constituents.

For single epoch kinematic PPP at few centimeters (e.g., 5-cm level), or millimeter-level static PPP over 24-hour period and/or for stations that are far from the oceans, ocean loading can be safely neglected [16]. However, when tropospheric or clock solutions are required, ocean loading must be taken into account even for 24-hour static point positioning processing, unless the station (or site) is far (>1000 km) from the nearest coastline.

Pole Tide

Pole tide is an additional tidal constituent due to the polar motion of the Earth (from the Chandler wobble with a period of 433 days ≈ 14 months). Because of this geophysical effect, it is necessary to take into account the resultant displacement on station coordinates derived from GNSS observations carried out over periods longer than two months [30].

From the 2010 IERS Conventions [22, p. 116], if X, Y, Z are the Cartesian coordinates of a station in a right-handed equatorial coordinate system, the changes in them due to polar motion are

$$\Delta\mathbf{r}_{pol} = [dX, dY, dZ]^T = R^T[S_\theta, S_\lambda, S_r]^T \tag{5.36}$$

where

$$R = \begin{pmatrix} cos\,\theta\ cos\,\lambda & cos\,\theta\ sin\,\lambda & -sin\,\theta \\ -sin\,\lambda & cos\,\lambda & 0 \\ sin\,\theta\ cos\,\lambda & sin\,\theta\ sin\,\lambda & cos\,\theta \end{pmatrix}$$

and

$$\begin{aligned} S_r &= -33sin2\theta(m_1\,cos\,\lambda + m_2\,sin\,\lambda), \\ S_\theta &= -9cos2\theta(m_1\,cos\,\lambda + m_2\,sin\,\lambda), \\ S_\lambda &= 9cos\theta(m_1\,sin\,\lambda - m_2\,cos\,\lambda), \end{aligned} \tag{5.37}$$

The radial displacement S_r and the horizontal displacements S_θ and S_λ (positive upward, south, and east, respectively, in a horizon system at the station) are in millimeters. The variables m_1 and m_2 in arcseconds represent the secular variation of the position of the Earth's mean rotation pole. The TRF coordinates of the secular variation are expressed in terms of the polar motion variables x_{pol} and y_{pol} (see, e.g., Chap. 3, Fig. 3.9) and their appropriate running averages $\bar{x}_{pol}$ and $-\bar{y}_{pol}$, such that:

$$m_1 = x_{pol} - \bar{x}_{pol}, \quad m_2 = -(y_{pol} - \bar{y}_{pol}) \tag{5.38}$$

See IERS Technical Note Number 36, page 115, for additional details on the coefficients of the IERS (2010) mean pole model [22].

References

1. Ashby, N. (2003). Relativity in the global positioning system. *Living Reviews in Relativity, 6*(1). https://doi.org/10.12942/lrr-2003-1
2. Ashby, N. (2004). The Sagnac effect in the global positioning system. In *Relativity in rotating frames*. ISBN: 978-1-40-201805-3.
3. Bilich, A., & Larson, K. M. (2009). It's not all bad: Understanding and using GNSS multipath. *GPS World, 20*(10), 31–39.
4. Bilich, A., & Mader, G. (2010, September 21–24). GNSS absolute antenna calibration at the National Geodetic Survey. In *Proc. of the 23rd Intl. Technical Meeting of the Satellite Division of The Institute of Navigation, Portland, OR, USA*.
5. Bilich, A., Schmitz, M., Görres, B., Zeimetz, P., Mader, G., & Wübbena, G. (2012, July 23–27). *Three-Method Absolute Antenna Calibration Comparison*. IGS Workshop University of Warmia and Mazury (UWM), Olsztyn, Poland.
6. Bohm, J., Niell, A., Tregoning, P., & Schuh, H. (2006). Global Mapping Function (GMF): A new empirical mapping function based on numerical weather model data. *Geophysical Research Letters, 33*(L07304). https://doi.org/10.1029/2005GL025546
7. Dong, D., Fang, P., Bock, Y., Cheng, M. K., & Miyazaki, S. (2002). Anatomy of apparent seasonal variations from GPS-derived site position time series. *Journal of Geophysical Research, 107*(B4). https://doi.org/10.1029/2001JB000573
8. EL-Hattab, A. I. (2013). Influence of GPS antenna phase center variation on precise positioning. *NRIAG Journal of Astronomy and Geophysics, 2*(2), 272–277. https://doi.org/10.1016/j.nrjag.2013.11.002
9. Hakansson, M. (2017). *Hardware Biases and Their Impact on GNSS Positioning*. Licentiate dissertation, Stockholm.
10. Håkansson, M. (2020). GNSS hardware biases in code and carrier phase observables. KTH School of Architecture and the Built Environment, Stockholm, Sweden. ISBN 978-91-7873-454-2.
11. Håkansson, M., Jensen, A. B. O., Horemuz, M., et al. (2017). Review of code and phase biases in multi-GNSS positioning. *GPS Solution 21*, 849–860.
12. Hernandez-Pajares, M., Juan, M., & Sanz, J. (2007). Second-order ionospheric term in GPS: Implementation and impact on geodetic estimates. *Journal of Geophysical Research, 112*, 1–16.
13. Hofmann-Wellenhof, B., Lichtenegger, H., & Wasle, E. (2008). *GNSS–Global Navigation Satellite Systems: GPS, GLONASS, Galileo & More*. Berlin: Springer. ISBN 978-3-211-73012-6.
14. Hoque, M., & Jakowsky, N. (2008). Estimate of higher order ionospheric errors in GNSS positioning. *Radio Science, 43*, 1–15.
15. Julien, O., Issler, J.-L., Lestarquit, L., Ries, L., & Hein, G. (2015, March/April). Estimating ionospheric delay using GPS/Galileo signals in the E5 Band. In *Inside GNSS* (pp. 54–64).
16. Kouba, J. (2009). A guide to using International GNSS Service (IGS) products. National Resources Canada.
17. Leick, A., Papoport, L., & Tatarnikov, D. (2015). *GPS satellite surveying* (4 ed.). Hoboken, NJ: John Wiley & Sons, Inc. ISBN 978-1-118-67557-1.

18. Mader, G. L. (1999). GPS antenna calibration at the National Geodetic Survey. *GPS Solutions, 3*(1), 50–58. https://doi.org/10.1007/PL00012780
19. Meindl, M. (2011). *Combined analysis of observations from different Global Navigation Satellite Systems* (Vol. 83). Geodätisch-geophysikalische Arbeiten in der Schweiz. Schweizerische Geodätische Kommission. ISBN: 978-3-908440-27-7.
20. Parkinson, B., Spilker, J., & Enge, P. (1996). *Global positioning system, theory and applications* (Vols I and II). Reston, VA, USA: American Institute of Aeronautics.
21. Paziewski, J., & Wielgosz, P. (2014). Accounting for Galileo–GPS intersystem biases in precise satellite positioning. *Journal of Geodesy, 89*(1), 81–93.
22. Petit, G., & Luzum, B. (Eds.) (2010). IERS Conventions (2010). In *IERS Technical Note 36* (179 pp.). IERS Conventions Centre, Frankfurt am Main, Germany, available at http://www.iers.org/TN36/. ISSN: 1019-4568.
23. Petovello, M. (2011, March/April). Clock offsets in GNSS receivers. In *Inside GNSS* (pp. 23–25).
24. Petrie, E. J., Hernandez-Pajares, M., Spalla, P., Moore, P., & King, M. A. (2011). A Review of higher order ionospheric refraction effects on dual frequency GPS. *Surveys in Geophysics, 32*(3), 197–253. https://doi.org/10.1007/s10712-010-9105-z
25. RINEX: The Receiver Independent Exchange Format Version 3.04. International GNSS Service (IGS), RINEX Working Group and Radio Technical Commission for Maritime Services Special Committee 104 (RTCM-SC104).
26. Rothacher, M., Gurtner, W., Schaer, S., Weber, R., Schluter, W., & Hase, H. O. (1996). Azimuth- and elevation-dependent phase center corrections for geodetic GPS antennas estimated from GPS Calibration campaigns. In G. Beutler et al. (Eds.), *GPS trends in precise terrestrial, airborne, and spaceborne applications, IAG Symposium* (Vol. 115, pp. 333–338). Springer.
27. Schmid, R., & Rothacher, M. (2003). Estimation of elevation-dependent satellite antenna phase center variations of GPS satellites. *Journal of Geodesy, 77*(7–8), 440–446. https://doi.org/10.1007/s00190-003-0339-0
28. Seeber, G. (2003). *Satellite geodesy*. New York: Walter de Gruyter.
29. Sleewaegen, J.-M., Simsky, A., De Wilde, W., Boon, F., & Willems, T. (2012, May/June). Demystifying GLONASS inter-frequency carrier phase biases. In *Inside GNSS* (pp. 57–61).
30. Subirana, S. J., Juan Zornoza, J. M., & Hernández-Pajares, M. (2013). *GNSS Data Processing: Vol I: Fundamentals and Algorithms*. ESA.
31. Teunissen, P. J. G., & Kleusberg, A. (Eds.) (1998). *GPS for geodesy* (2nd ed.). Berlin, Germany: Springer. ISBN 3-540-63661-7.
32. Teunissen, P., & Montenbruck, O. (2017). *Springer handbook of global navigation satellite systems*. Springer. ISBN 978-3-31-942926-7.
33. Wanninger, L. (2011). Carrier-phase inter-frequency biases of GLONASS receivers. *Journal of Geodesy*. https://doi.org/10.1007/s00190-011-0502-y
34. Wu, J., Wu, S., Hajj, G., Bertiguer, W., & Lichten, S. (1993). Effects of Antenna orientation on GPS carrier phase measurements. *Manuscripta Geodaetica, 18*, 91–98.

Chapter 6
GNSS CORS Networks and Data

6.1 Geodetic CORS

6.1.1 Definitions

CORS is the acronym for Continuously Operating Reference Station which has been set up to permanently track GNSS signals continuously. Additional tasks include storing the raw data and in some cases processing the data and transmitting (broadcasting) to rover stations (receivers).

The basic design of a GNSS CORS system mainly comprises (1) a geodetic quality antenna affixed to a permanent structure or monument (e.g., as shown in Fig. 6.1) with some known physical reference (measurement point), (2) a geodetic quality receiver (e.g., as shown in Fig. 6.2), and (3) continuous power supply and communication infrastructure such as cables and telemetry (e.g., as shown in Fig. 6.3). A threaded antenna mount (as shown in Fig. 6.4) typically exists between the monument (or structure) and the antenna, with the physical reference (measurement point) being located on the top surface of the mount to facilitate measurement of ARP offset.

Therefore, as a minimum, each geodetic CORS requires a geodetic quality receiver, an antenna installed on a stable monument or structure, communications, and a power supply. In some cases, a computer device is also installed for data transmission and control. Most geodetic receivers for CORS systems have suitable communication ports and management software with the ability to stream raw data back to a central server location and a supplementary configuration for back up and reliability. A user interface is required to configure and maintain the system and settings, and this may be done remotely, for example, by Internet connection or radio communication.

Many factors go into the installation and operation of GNSS CORS, based upon best practices and guidelines [25, 28]. For example, site location, site stability, satellite visibility, equipment security, and proximity to power and communications

C. A. Ogaja, *Introduction to GNSS Geodesy*,
https://doi.org/10.1007/978-3-030-91821-7_6

Fig. 6.1 GNSS CORS antenna on a monument pillar

are some of the important considerations when installing a geodetic CORS. Other considerations include, for example, installing ground-based monuments versus structure- or roof-based monuments. Some of the operational considerations also include the following:

1. **Communications and data access** depending on whether the CORS stores the observed raw data for download and post-processing or processes the data and transmits to rover receivers (e.g., through subscription services). In the latter case, the CORS often operates as part of a network that is capable of estimating and resolving carrier-phase ambiguities in what is often referred to as a network solution.
2. **Uninterruptible power supply** for supporting the continuous operation of the CORS system and, if necessary, additional power supply infrastructure such as solar panels for supporting the system's continuous operation, but especially in the event of a power outage.
3. **Site security** (such as security fence in the case of ground-based monuments, locked box enclosure and/or building enclosure) for protecting the CORS site and equipment from physical disturbance, vandalism, theft, weather elements, lightning, animals, birds, and insects.

Fig. 6.2 GNSS CORS receiver in enclosure box/cabinet

4. **Consistent ARP** for relating the CORS antenna measurements to the physical reference mark (point) on the monument, through antenna models such as the NGS-validated phase center variation models and/or equipment manufacturer-supplied models.

6.1.2 Guidelines

Many standard documents outline the best practices and guidelines for establishing geodetic CORS. Such documents are authored and maintained by leading geodetic organizations and governmental authorities such as the International GNSS Service (IGS) [20], US National Geodetic Survey (NGS) [23, 25], Natural Resources Canada (NRCan) [28], Australian and New Zealand Intergovernmental Committee on Surveying and Mapping (ICSM) [17], and many others. The following are summaries of some of the recommended practices and guidelines as found in those documents.

Fig. 6.3 GNSS CORS power supply and telemetry

Fig. 6.4 Antenna mount without a GNSS antenna

Site Location The CORS site location should be on a stable ground or structure, with good sky visibility, without RF or multipath sources in close proximity, and the ARP should be such that it can maintain a fixed position in three dimensions relative to the ground surface. In the case of a constructed monument, increasing the depth of the monument improves its stability. The CORS antenna and monument must be well anchored to the ground to ensure that the position and velocity associated with the CORS site represent the crustal position and velocity and not that of the antenna. Therefore, it is important to beware of, and to minimize, the impact of undesirable site conditions such as ground heaving, swelling or shrinking, soil expansion and contraction, slope instability, active fluid or gas pumping, sink holes, mines, and monument and structural thermal expansion and contraction.

Satellite Visibility The CORS should be located in an area with minimum or no traffic, minimal obstructions, and minimum likelihood of change in the surroundings such as future tree or shrub growth and building or structural additions. The location should be such that, for example, there are no obstructions above 10 degrees of the antenna reference point (ARP) horizon, and minimal obstructions otherwise.

Survey Mark and Antenna Eccentricity The definition of (a) *CORS monument* as the "structure (e.g., pillar, building, etc.), including the mount, which keeps the GNSS antenna attached to Earth's surface," (b) *Survey mark* (measurement point) as "a unique and permanent point on the monument to which the antenna reference point (ARP) is measured," and "must remain invariant with respect to the monument," and (c) *Antenna eccentricity* (also known as the ARP offset) as "the vertical and horizontal distances from the mark to the ARP."

Radio Frequency Environment Every effort should be made to avoid proximity to radio frequency equipment (such as TV stations, microwave, FM radio stations, cellular towers, VHF and UHF repeaters, RADAR) and high-voltage power lines. Such radio frequency sources and high-voltage power lines can cause additional noise, interference, and loss of lock to the GNSS signals received by the CORS antenna or even render the CORS inoperable.

Ground-Based Monument Pillar monument (Fig. 6.1) with a deep foundation for stability and tall enough to enhance satellite visibility. For example, NGS guidelines specifically recommend that pillar monuments "be approximately 1.5 m above the ground surface to mimic the geometry used at NGS's antenna phase center calibration facility." In addition, NGS has specific requirements on the size of the top of the pillar in comparison to the antenna size [23], as well as recommendations on the placement of antenna to mitigate multipath issues. Braced monuments are well anchored to the ground and stable, but generally more expensive than pillar monuments.

Roof-Based Monument A roof-based monument should take into account factors such as type and age of building, its condition, height, and roof type. For example, solid brick or reinforced concrete buildings are recommended, and older buildings increase the likelihood that all primary settlings have occurred. High-rise buildings are recommended for better satellite visibility. Non-metal roofs and non-metal-walled buildings would minimize the effects of thermal expansion and multipath issues. The monument material type, location, and attachment to the building also matter (e.g., stainless steel is recommended for longevity, while aluminum is not due to higher thermal expansion, and "the mount must be bolted directly to the main part of the building; a load-bearing wall near a corner is recommended" [23]).

Equipment and Firmware Upgrades Equipment should be inspected regularly for damage and deterioration, and firmware must be kept updated. It is also recommended that the equipment be upgraded or replaced as the technology changes (e.g., new GNSS signals added). Otherwise equipment changes should be minimized as they have the potential to cause changes in the CORS positions.

Receiver and Antenna Settings The CORS receiver and antenna should be supported by manufacturers and capable of at least dual-frequency (e.g., L1 and L2) tracking, in addition to the following capability and settings: (1) track additional GPS and GNSS signals, (2) set to track all satellites down to zero degree elevation (strongly recommended), (3) record raw data at 30-, 15-, 5-, or 1-second sampling intervals, (4) create hourly sessions (strongly preferred), or 24-hr sessions of GPS time, and (5) track all satellites regardless of health status.

6.2 Tracking Networks and Services

6.2.1 Introduction

A CORS network consists of several stations (CORSes) and can be classified as either offline or online (real-time) network. In the case of an offline network, archived data is typically available to the user for post-processing applications using a standard format such as RINEX (Sect. 6.3.1). An online network is a cluster of CORSes interconnected by reliable communications to enable real-time computations, control, and transmissions. RTCM format (Sect. 6.3.2) is normally used for real-time transmission of GNSS corrections from a real-time network to the user's receiver (rover station).

Real-time CORS networks are more widespread in densely inhabited areas, especially for local or regional coverage, while the post-processing variant is more applicable for expansive coverage and places with less population density. A real-time network typically has shorter inter-station distances compared to the post-processing variant, mainly due to the differences in data processing strategies and requirements for handling station-dependent biases. For example, in a real-time network, shorter inter-station distances (e.g., <20 km) imply that network data processing can utilize short duration data (e.g., 1 second epochs) to handle such biases. In a post-processing network, longer datasets (such as hourly or 24-hour files) are necessary for modeling and removing station-dependent biases and error sources such as atmospheric and Earth deformation effects. Such biases cannot be eliminated with differenced observables (Chap. 5, Sect. 5.3.1) when longer inter-station distances (e.g., up to 1000 km) are involved, but can be better handled with empirical models using long observation datasets.

CORS networks can also be differentiated on the basis of certain criteria or factors such as site or monument stability, primary purpose (and applications), inter-station spacing, equipment quality, compliance with specific site guidelines (e.g., [20]), and so forth. For example, Australia's ICSM guidelines version 2.1 of 2014 [17] specifies a hierarchy or tier structure for differentiating between CORS networks in Australia, based on inter-station distances, as originally proposed by [32].

In general, the main advantages of CORS networks include the reduction or elimination of baseline dependent errors, automatic referencing of GNSS measurements and results to reference frames and geodetic datums (see Chap. 3 "Reference Systems in GNSS Geodesy"), and savings[1] for the user who would otherwise need to invest on a GNSS reference receiver (or station) and the time for setup and/or installation.

6.2.2 *IGS Network*

Figure 6.5 shows the IGS network in 2021. It is a collection of CORS stations operated by many different organizations pooling their resources under the IGS umbrella. The IGS (www.igs.org) was founded in 1994 as a voluntary federation of self-funding agencies, universities, and research institutions in more than 100 countries, working together to (1) provide "the highest precision GPS satellite orbits in the world," (2) provide "free and open access to the highest precision products available for scientific advancement and public benefit," (3) produce "products that support realization of the International Terrestrial Reference Frame while providing access to tracking data from over 500 worldwide reference stations," (4) work "for the continuous development of new applications and products through Working Groups and Pilot Projects," and (5) support "geodetic research and scholarly publications."

The IGS global network of continuously operating geodetic quality stations tracks all the different GNSS constellations including GPS, GLONASS, Galileo, BeiDou, QZSS, and SBAS. The operational structure of IGS [5] includes components such as Data Centers, Analysis Centers, the Central Bureau, the Governing Board and Associate Members, as well as Pilot Projects and Working Groups. Raw station data are archived at the IGS Global Data Centers and multiple Regional Data

[1] CORS networks benefit the users by utilizing one GPS receiver as the operation of the reference station is performed by the service provider of the CORS network.

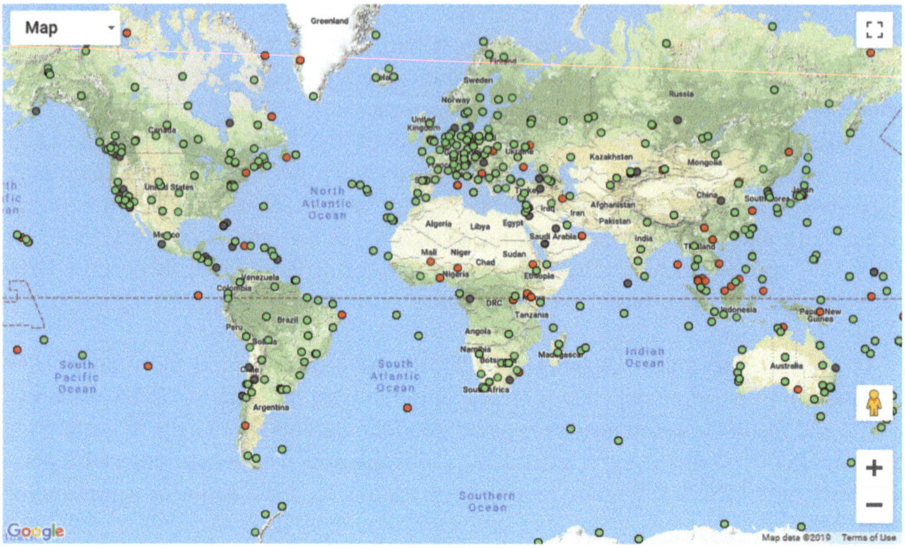

Fig. 6.5 IGS CORS network [19]

Centers. Analysis Centers regularly process the data and contribute products to the Analysis Center Coordinator, who produces the official IGS combined products.

The IGS data products include tracking station coordinates and velocities, satellite orbits, satellite and tracking station clock information, zenith tropospheric path delay estimates, global ionosphere maps, and the Earth's rotation parameters. These products are in support of other efforts such as:

- Improving and extending the ITRF maintained by the IERS
- Monitoring Earth deformations
- Monitoring the troposphere and ionosphere
- Orbit determination for non-GNSS scientific satellites
- Earth rotation monitoring

The IGS Real-Time Service is a GNSS orbit and clock correction service that enables Precise Point Positioning (PPP) at worldwide scales. The products from this service enable applications such as hazard detection and warning, weather forecasting, time synchronization, geophysical monitoring, imagery control, and many other public benefit applications.

6.2.3 *NOAA Network*

The US NOAA's National Geodetic Survey (NGS) manages a multipurpose cooperative CORS network involving government, academic, and private organizations.

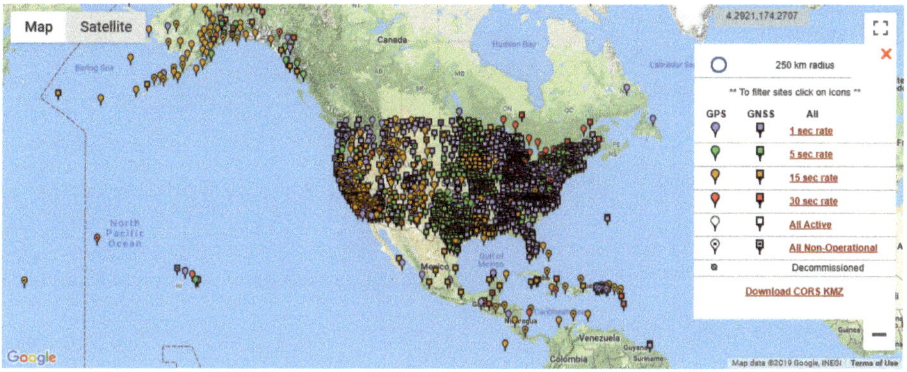

Fig. 6.6 NOAA CORS network [24]

The sites are independently owned and operated, and each agency shares their data with NGS, and NGS in turn analyzes and distributes the data free of charge. Figure 6.6 shows the NGS network in 2021.

The primary objective of the NGS CORS network is to enable GNSS users such as surveyors and mapping professionals have access to the US National Spatial Reference System (NSRS[2]) to ensure that their positional coordinates are compatible with those determined by others. The NGS CORS network also contributes to the determination of ITRF through a specific set of stations that must be operated at the highest standards and federally owned to ensure consistency and longevity. Although the entire NGS network is a cooperative network from over 200 government, academic, and private organizations, only a limited set of stations (known as the NOAA Foundation CORS Network) across the USA provide access to the NSRS. The network data are submitted for inclusion as part of the IGS network which subsequently contributes to the ITRF and plate rotation models.

6.2.4 APREF Network

The Asia–Pacific Reference Frame (APREF) network (comprised of permanent tracking stations covering the Asia and Pacific region and its surroundings) was the outcome of a project started by member countries following the United Nations Regional Cartographic Conference (UNRCC) for Asia and the Pacific, 26–29 October 2009, in Bangkok, Thailand [7]. The goal was to establish a regional geodetic reference frame for the Asia–Pacific region, including countries such as Afghanistan, Australia, China, Fiji, Indonesia, Iran, Iraq, Japan, Malaysia, Mon-

[2] NSRS is a consistent coordinate system that defines latitude, longitude, height, scale, gravity, orientation, and shoreline throughout the United States.

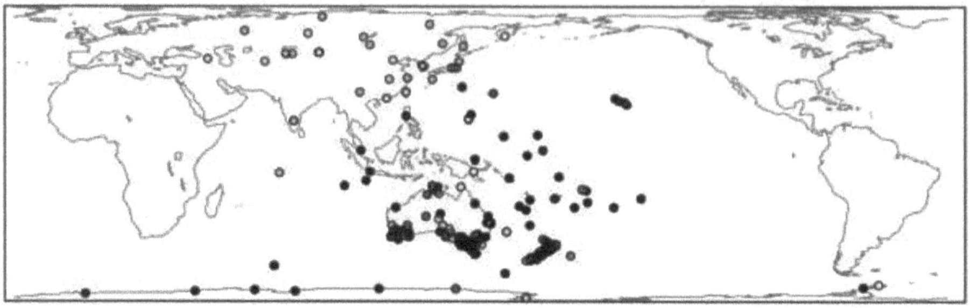

Fig. 6.7 APREF CORS network [16]

golia, Nauru, New Zealand, Philippines, Samoa, Solomon Islands, Tonga, Tuvalu, Vanuatu, and pertinent others. One of the primary short-term goals was to encourage sharing of GNSS data from CORSes in the region and to develop an authoritative source for their coordinates and velocities. The long-term objectives were to (1) maintain a permanent CORS network for the region, in close cooperation with the IGS, (2) contribute to the ITRF, and (3) establish "a dense velocity field model in Asia and the Pacific for scientific applications and the long-term maintenance of the Asia-Pacific reference frame" [7]. Figure 6.7 shows the geographic distribution of stations contributing to the network.

The APREF's organizational structure is in some ways similar to the IGS (e.g., including a Central Bureau, Data Centers, and Analysis Centers) and follows the IGS station, data, and analysis standards. APREF was mandated by Resolution 1 (Regional Geodesy) of the 18th UNRCC-AP, and also endorsed by the IGS, United Nations Office for Outer Space Affairs (UNOOSA), and the Federation of International Surveyors (FIG).

The APREF CORS network is a voluntary network and participating organizations and agencies are encouraged to share (contribute) data from their stations, allow user access to their online archived data and products, and routinely participate in the combination analyses for the combined APREF solution of station coordinate and velocity estimates. This in turn contributes to the densification of the ITRF in the Asia and Pacific region [15].

6.2.5 EUREF Network

EUREF (European Reference Frame) network (Fig. 6.8) is a network of permanent CORS covering the European continent [2, 21, 38]. In addition to publicly available, continuously archived raw GNSS tracking data, the data products of the network

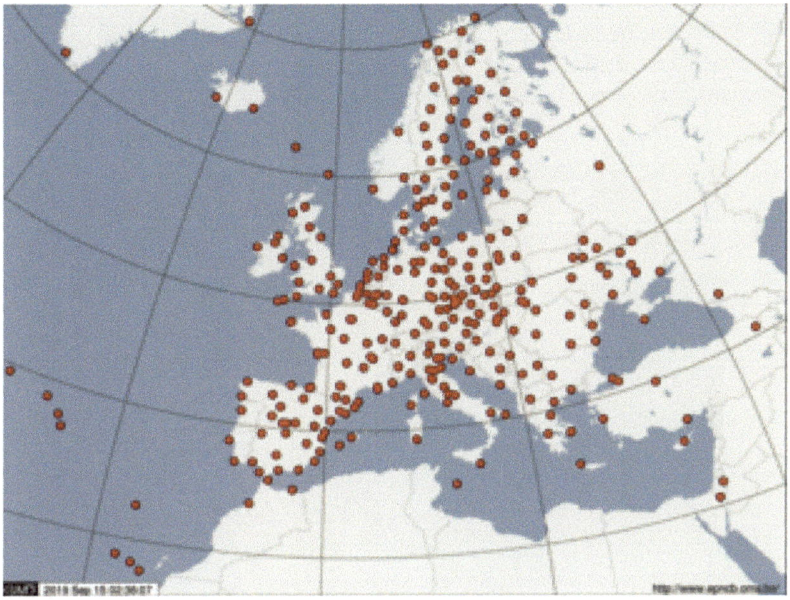

Fig. 6.8 EUREF CORS network [2]

include station positions and velocities in the ITRF and ETRS89.[3] It is a voluntary network with contributions from over 100 European agencies and universities, and the operational structure includes *data centers* providing access to station tracking data,[4] *analysis centers*, and a *Central Bureau* that oversees the management.

The network is operated under the umbrella of the IAG Regional Reference Frame sub-commission for Europe, which was founded in 1987 at the IUGG General Assembly in Vancouver. As is common with other continental and/or regional networks, the EUREF network is first and foremost applied to produce coordinates in a reference system (ETRS89) which is tied to a stable European plate (i.e., "the relations between European stations are kept fixed"[5]). The network's primary objective is a stable regional reference frame for the European continent in which some of the network stations also contribute to the global reference frame, ITRF.

[3] "The primary purpose of the EUREF network is to provide access to the European Terrestrial Reference System 89 (ETRS89) which is the standard precise GNSS coordinate system throughout Europe." (http://epncb.eu/).

[4] CORS station tracking data typically includes phase and pseudorange observations, satellite ephemeris, and meteorological sensor data where available.

[5] In the ITRF (the global reference frame), "plate tectonics cause the coordinates of European stations to slowly change in the order of about 2.5 cm/year".

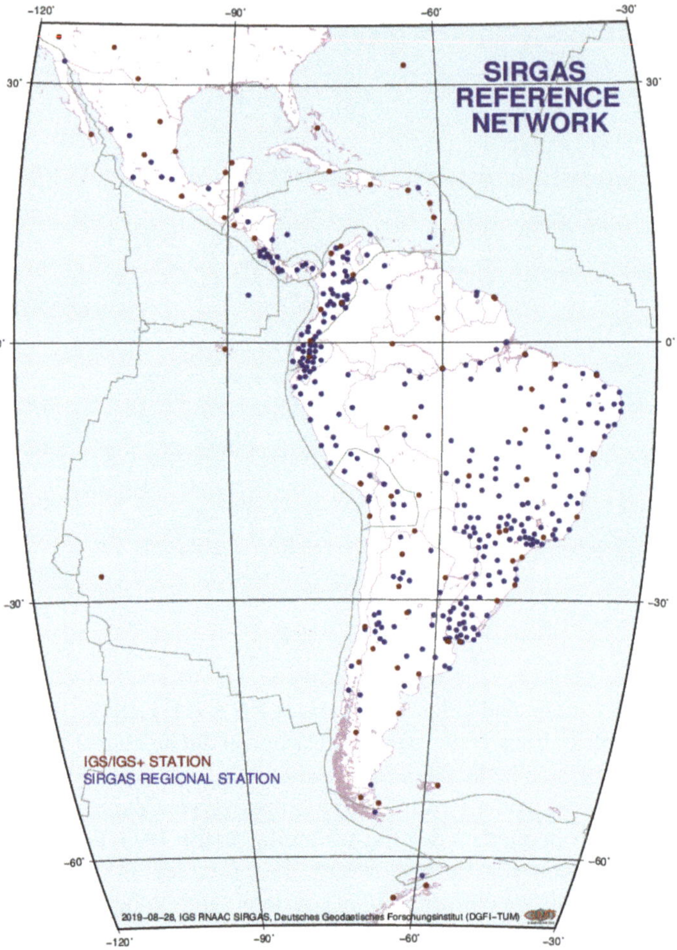

Fig. 6.9 SIRGAS CORS network (www.sirgas.org) [4]

6.2.6 SIRGAS Network

SIRGAS (Spanish acronym for Geocentric Reference System for the Americas) network is a network of CORS sites distributed over Latin America as shown in Fig. 6.9. The operational structure of SIRGAS is based on the contribution of more than 50 organizations, which install and operate their own stations and voluntarily provide the tracking data for routine processing of the network [4].

The SIRGAS also has its foundations in the UN resolution (A/RES/69/266) "A Global Geodetic Reference Frame (GGRF) for Sustainable Development," and also as a member of the IAG Sub-commission 1.3 which deals with the definitions and realizations of regional reference frames and their connection to the global International Terrestrial Reference, ITRF.

6.2.7 AFREF Network

The African Reference Frame (AFREF) project was launched in the year 2000 as one of the IAG sub-commission 1.3 (Regional Reference Frames) projects, with the aim of establishing a unified CORS network for Africa. Although the goal was not yet fully accomplished at the time of this writing, an active web server at http://www.afrefdata.org/ maintained continuous log files for the participating reference stations around the continent.

As of 2021, the project did not yet have an established operational structure (e.g., a Central Bureau and Data Centers for routine data processing) like the other regional networks. There were continuing efforts toward the establishment of such an operational structure, but several challenges remained. For example, nearly all African countries supported and were willing to implement the project but found it difficult to include the project among their development priorities due to economic difficulties [3, 40].

Most of the CORS (over 100 in 2021) that have been established in Africa are through the efforts of international organizations such as IGS, NOAA's NGS, UNAVCO's AfricaArray, and SEGAL (Space and Earth Geodetic Analysis Laboratory) in Portugal (http://segal.ubi.pt). Some countries have also established their own CORS networks, for example, South Africa's TrigNet (>55 CORS), RwandaGeonet by Rwanda' Natural Resources Authority (RNRA), BotswanaNet (>55 CORS), and Nigeria's NIGNET. Regional institutions such as RCMRD (www.rcmrd.org) [31] are actively involved in promoting the AFREF project through collaborations with regional and international partners as well as offering of workshops and training (on AFREF and GNSS data processing) for the development of needed manpower.

6.2.8 Other Networks

There are various other categories of CORS infrastructure that exist around the world. These include, for instance:

- Private subscription networks by GNSS equipment vendors and manufacturers such as Trimble, Hexagon, and Topcon. These entities operate some of the world's largest networks. For example, Trimble operates over 6500 base stations throughout the world, while Hexagon's HxGN SmartNet (hxgnsmartnet.com) operates over 4500 base stations worldwide. Topcon also operates a network of base stations covering the United States as well as parts of Europe, Australasia, and the Middle East.

- Real-time monitoring networks such as Japan's GEONET and UNAVCO's Plate Boundary Observatory (PBO) network. These kinds of networks are focused on science and monitoring of geohazards such as earthquakes, plate tectonics, and other forms of Earth deformation.
- Networks owned by local governments and departments. For example, in the United States, networks owned and operated by state Departments of Transportation are commonplace. These networks are mostly installed by GNSS equipment vendors through contracts and operate as part of local government infrastructure for various tasks and projects such as land surveys and highways construction and maintenance.
- Networks and services by private individuals and small firms or organizations where sharing of information or data is mostly profit-driven. In this category, the CORS owners are under no requirement to follow guidelines except if/when they opt to participate in certain joint ventures such as regional Reference Frame projects (Sect. 6.2.2–6.2.7).

It should also be noted that some of the services overlap in terms of data sharing and governance. For example, some of the UNAVCO's PBO stations contribute to the IGS network (Sect. 6.2.2).

6.3 GNSS Raw Data Exchange Formats

The GNSS industry is such that each receiver type has its own proprietary format for storing data collected from GNSS observations. Consequently, simultaneously tracked data from different receiver types cannot be easily processed with one particular software package. This poses a challenge, especially for multiple data collected from tracking networks that require combined processing. The different networks worldwide (Sect. 6.2) are joint efforts between different stakeholders, and equipment types by different manufacturers are involved. But even small projects or a single network by one stakeholder may have more than one receiver type collecting raw data.

In addition to storing data for post-processing, some networks (as discussed in Sect. 6.2.8) provide subscription services for real-time streaming of data to clients. This causes similar challenge since in most cases the subscribed user would not necessarily have the same receiver type as the ones being used by the network to provide the service. The analogy can also be extended to a simple standalone project utilizing two different receiver types.

One way to solve the problem of dealing with different proprietary binary data formats is to define acceptable common raw data exchange formats with the following desired features: (1) non-proprietary and accessible to all, (2) can be used as interface between different receiver types, and (3) can be used by different data processing software systems.

The following sections discuss the available data exchange formats that have been developed to solve this problem.

6.3.1 RINEX Format

RINEX (Receiver INdependent EXchange) is the standard format for sharing raw GNSS data collected by different receiver types. Typically, each receiver type has its own proprietary binary format for storing observations from the satellites. Such data can be post-processed using software specially developed for a receiver by manufacturer (usually sold in a package with the receiver). However, for CORS networks and GNSS projects that use different receiver types by different manufacturers, it becomes problematic to process a variety of format, data, and file types collected in a given session.[6] One manufacturer's software would not process data from another manufacturer's receiver. For this reason, RINEX was created and adopted, starting in 1989 [18], as the standard ASCII file format into which the proprietary binary format data from any geodetic receiver can be converted for exchange and processing. Since then, the RINEX format has evolved through a number of versions, as shown in Table 6.1 in reverse chronological order.

Three different types of RINEX files have been used since the introduction of RINEX 2. These include observation files, broadcast navigation message files, and meteorological data files. The original RINEX file naming convention (used up to RINEX 2.11) was implemented in the MS-DOS era when file names were restricted to 8.3 characters. Modern operating systems typically support 255 character file names, and hence new file naming conventions have emerged that are more descriptive, flexible, and extensible.

RINEX Observation Files

Observation files contain specific observables that are defined for the geodetic processing software to interpret. These include:

- **Pseudorange (PR) or code**. The difference between receiver clock time at signal reception (in the time frame of the receiver clock) and the satellite clock time at signal transmission (in the time frame of the satellite clock). Due to the two different time frames, the computed time difference differs from the actual travel time by the receiver and satellite clock offsets from the reference time scale (typically GPS time). Thus,

  PR (in sec.) = travel time +
  receiver clock offset−satellite clock offset + other biases.

  PR (in m.) = actual range +
  $c \times$ (receiver clock offset−satellite clock offset + other biases)

[6] GNSS data session is the time period during which all the receivers are collecting data simultaneously.

Table 6.1 Summary of RINEX versions

RINEX version	Author
RINEX v. 3.05 – Released 01-Dec-2020 as the standard for exchange of GNSS data in RINEX 3 format – Supports all publicly available signals from GPS, GLONASS, Galileo, BeiDou, QZSS, and IRNSS constellations – Contains updates to fully support BDS II and BDS III, and adds missing flags and values to the GLONASS navigation message	IGS and RTCM-SC104
RINEX v. 3.04 – Released 23-Nov-2018 as the standard for exchange of GNSS data in RINEX 3 format (together with RINEX 3.02 and 3.03) – Supports all publicly available signals from GPS, GLONASS, Galileo, BeiDou, QZSS, and IRNSS constellations – Contains updates to support planned GLONASS CDMA signals, new BeiDou III, and QZSS II signals	IGS and RTCM-SC104
RINEX v. 3.03 – Released 14-Jul-2015 as the standard for the exchange of GNSS data in the RINEX 3 format (together with RINEX 3.02) – Added support for the Indian Regional Satellite System (IRNSS) and clarified several implementation issues in RINEX 3.02	IGS and RTCM-SC104
RINEX v. 3.02 – Released 03-Apr-2013 as the new standard for exchange of GNSS data in the RINEX 3 format – Introduced a new RINEX file naming convention – Enhanced 3.01 to include a new header message to specify the GLONASS code-phase bias, added support for QZSS and additional information concerning BeiDou	IGS and RTCM-SC104

(continued)

Table 6.1 (continued)

RINEX version	Author
RINEX v. 3.01 – Released on a test mode, leading to a new Version 3. Includes the unofficial v. 2.20 definitions for space-borne receivers. – Addressed the requirement to generate consistent phase observations across different tracking modes or channels, i.e., to apply 1/4-cycle shifts prior to RINEX file generation, if necessary, to facilitate the processing of such data – File types include GPS+GLONASS+Galileo observation files, broadcast navigation message files, and meteorological data files. Considered experimental by IGS and EUREF	W. Gurtner, AIUB, Switzerland and L. Estey, UNAVCO, Boulder, Colorado, USA[a]
RINEX v. 2.11 – Released 10-Dec-2007 as the new standard for exchange of GNSS data in the RINEX 2 format. – File types include GPS+GLONASS observation files, navigation message files, and meteorological data files. – C2, L2C/L5 and Galileo codes introduced. – Includes the definition of a two-character observation code for L2C pseudoranges.[11]	W. Gurtner, AIUB, Switzerland and L. Estey, UNAVCO, Boulder, Colorado, USA
RINEX v. 2.10 – A RINEX 2 subversion, file types include GPS+GLONASS observation files, navigation and meteorological data files. – Among other minor changes, allowing for sampling rates other than integer seconds and including raw signal strengths as new observables [9]	W. Gurtner, AIUB, Switzerland
RINEX v. 2 – Released Sept/Oct 1990 as the standard for exchange of GPS data in the RINEX format – Three file types defined: observation files, broadcast navigation message files, and meteorological data files [10] – Presented at and accepted by the Second International Symposium of Precise Positioning with the Global Positioning System in Ottawa, 1990, mainly adding the possibility to include tracking data from different satellite systems (GLONASS, SBAS) [18] and [8, 10]	W. Gurtner, AIUB, Switzerland

(continued)

Table 6.1 (continued)

RINEX version	Author
RINEX v. 1 – A draft version of an exchange format developed at the Astronomical Institute of the University of Bern for exchange of GPS data collected during the pan-European GPS campaign (EUREF) in May 1989 involving about 60 GPS receivers from four different manufacturers [8]. Original version 1 presented at and accepted by the 5th International Geodetic Symposium on Satellite Positioning in Las Cruces, 1989.[12, 18].	W. Gurtner, AIUB, Switzerland

[a] Werner Gurtner, Astronomical Institute of the University of Bern, Switzerland and Lou Estey, UNAVCO, Boulder, Colorado, USA

where c is the speed of light. The observables are stored in units of meters [8], and hence the second expression above applies. The original RINEX 2 format has three specific PR observation codes: C1 (from C/A code measurement on L1 frequency), P1, and P2 (from P-code measurements on L1 and L2, respectively), which was primarily for the GPS constellation but also GLONASS. However, new satellite constellations (Galileo, BeiDou, etc.) and signals (e.g., L5) led to the modification of observation codes as shown in the RINEX 2.11 document [11] "10.1.1 New Observation Codes." New features were introduced for RINEX 3 due to the new signal structures for GPS, Galileo, and BDS that make it possible to generate code and phase observations based on one or a combination of several channels. The observation codes increased from two (versions 1 and 2) to three characters by adding a signal generation attribute, e.g., as detailed in pp. 16–22 of the RINEX Version 3.04 document [18]. Table 3 of the document [18] defines the components of the new observation codes.

- **Phase observable**. Shown with observation code L# in a Rinex 2.11 and L## in a Rinex 3.xx file can be generally explained using

$$L(t) = N + \phi(t_0) + \Delta\phi(t) = N + \phi(t) + I(t) \tag{6.1}$$

where t is the epoch of observation, t_0 is the start or initial epoch, N (unknown constant,[7] but initially "guessed"[8]) is the **integer** (whole) number of cycles

[7] "Switching on a receiver at epoch t_0, the instantaneous fractional beat phase is measured. The initial integer number N of cycles between satellite and receiver is unknown. However, when tracking is continued without loss of lock, the number N, also called integer ambiguity, remains the same", p. 107 of [14].

[8] An estimate of N can be obtained by one of several means, for example, from the pseudorange measurement or Doppler (see, e.g., p. 36–38 of [30]).

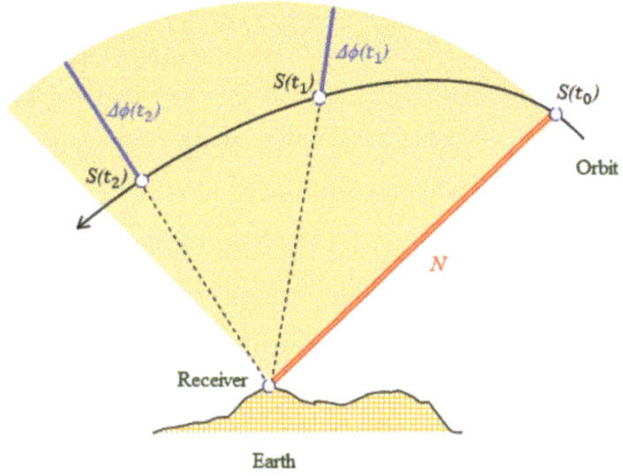

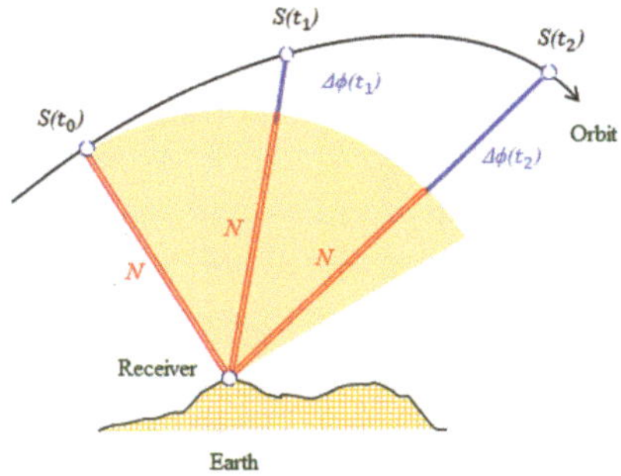

Fig. 6.10 Geometry of phase observable for approaching (A) and departing (B) satellite. For simplicity, the initial phase $\phi(t_0)$ (Eq. 6.1) is assumed to be zero. N is unknown but initially estimated (and remains constant unless (or until) there is a loss of lock on signal)

between satellite and receiver at the initial epoch, $\phi(t_0)$ is the phase (fractional cycle)[9] reading at the initial epoch, and $\Delta\phi(t)$ is the cumulative count of change in phase since the initial epoch (see Fig. 6.10).[10] At the initial epoch, $\Delta\phi(t_0)$ is zero, and thus N (**the phase ambiguity**) is the only unknown in a $L(t_0) = N + \phi(t_0)$. After some duration of continuous measurement, $\Delta\phi(t)$ would accumulate to a real-numbered value so that the measured fractional phase $\phi(t)$ at epoch t is augmented by I, the number of integer cycles since the initial epoch t_0. Thus, from Eq. 6.1, it can be explained that

$$\phi(t) = frac[\phi(t_0) + \Delta\phi(t)] \tag{6.2}$$

$$I(t) = INT[\phi(t_0) + \Delta\phi(t)] \tag{6.3}$$

The phase observation must be tracked continuously. Loss of lock on phase causes a jump of an **integer** (full) number of cycles in the phase data ("cycle slip"), hence a different N value on new phase lock or initialization. Such occurrences of cycle slips (and gaps) in observables should be detected and repaired by processing software. Slightly different procedures are usually applied multiple times to find all of the cycle slips (and gaps). The phase observables reported in RINEX are not corrected for external effects such as atmospheric refraction and satellite clock offsets.

---

[9] $\phi(t_0)$ is equivalent to $(\phi_{rcv} - \phi^{sat})$ in Eq. 5.4 in Chap. 5.

[10] Ambiguity only occurs at the initial epoch. It is the unknown number of full cycles at the initial epoch. Once the receiver gets a lock on a satellite (2nd epoch onward) it can count the number of full cycles. See, e.g., [30].

- **Time**. Generally defined as the time of observation, expressed either in the time frame of the satellite clock at signal transmission or in the time frame of the receiver clock at signal reception. In the RINEX 3 format [18], it is defined as the receiver time of the received signals and is identical for both phase and pseudorange (code) measurements and for all satellites observed at the epoch. For single system data files (i.e., GPS or GLONASS, or Galileo, or BDS), the time is by default expressed in the time frame of the respective satellite system.

In a case where the receiver or a conversion software adjusts the measurements using real-time-derived receiver **clock offsets**, such correction should be applied to all the three observables (pseudorange, phase, and epoch (time)), for example, as shown in Table 1 of [18].

Additionally, it should be noted for clarification how the phase observable is modeled into the phase observation equation (e.g., Eq. 5.4, Chap. 5). For epoch t of observation (where t applies to all terms except N and $\phi(t_0)$),

$$\begin{aligned}\lambda(N + \phi(t_0) + \Delta\phi) = |(\boldsymbol{P} + \boldsymbol{E} + \boldsymbol{O}) - (\boldsymbol{p} + \boldsymbol{e} + \boldsymbol{o})| + d_{trop} - d_{ion} \\ + d_{orb} + c\delta t_{rcv} - c\delta t^{sat} + cB - cb + \lambda W + \gamma + M + v\end{aligned} \tag{6.4}$$

$$\begin{aligned}\lambda\Delta\phi = |(\boldsymbol{P} + \boldsymbol{E} + \boldsymbol{O}) - (\boldsymbol{p} + \boldsymbol{e} + \boldsymbol{o})| + d_{trop} - d_{ion} + d_{orb} + c\delta t_{rcv} \\ - c\delta t^{sat} + cB - cb - \lambda\phi(t_0) - \lambda N + \lambda W + \gamma + M + v\end{aligned} \tag{6.5}$$

$$\begin{aligned}\Phi = |(\boldsymbol{P} + \boldsymbol{E} + \boldsymbol{O}) - (\boldsymbol{p} + \boldsymbol{e} + \boldsymbol{o})| + d_{trop} - d_{ion} + d_{orb} + c\delta t_{rcv} \\ - c\delta t^{sat} + cB - cb + \lambda(\phi_{rcv} - \phi^{sat}) - \lambda N + \lambda W + \gamma + M + v\end{aligned} \tag{6.6}$$

where $\Phi = \lambda\Delta\phi$ and $\lambda(\phi_{rcv} - \phi^{sat}) = -\lambda\phi(t_0)$. All the other terms in the above equations are as explained in Eq. 5.4 in Chap. 5.

RINEX Navigation Message Files

RINEX navigation message files may contain navigation messages of more than one satellite system. The files contain satellite epoch and clock parameters (coefficients of SV clock polynomial), broadcast orbit/ephemerides data (i.e., *Keplerian elements plus correction coefficients for GPS, Galileo, BeiDou; position–velocity–acceleration vectors, aka state vectors, for GLONASS*)[11] for the satellites tracked, SV accuracy (i.e., expected accuracy of pseudorange measurements), and SV health (information on the health of the satellites). A single RINEX navigation message file contains all the navigation messages collected by the receiver during a particular time period (session). The file session is usually the same session as that covered by the corresponding observation file. For ease of post-processing and to increase temporal coverage, a composite navigation file can be created from all the receivers

[11] Broadcast orbits (aka ephemerides data) for GPS, Galileo, and BeiDou are in the format of Keplerian elements and correction coefficients. However, for GLONASS, they are in the format of state vectors (position-velocity–acceleration vectors).

in a campaign, to contain all nonredundant messages collected during the session. Optionally, the navigation message file may contain additional data in the header section, such as ionospheric correction parameters, time system correction terms (difference between GNSS system time and UTC or other time systems), and leap seconds [18, pp. A19–A21].

RINEX Meteorological Data Files
Meteorological data file contains data from session corresponding to that of the observation file. In the RINEX 3 format, it contains time-tagged pressure (in mbar), dry air temperature (in degrees Celsius), relative humidity (percent), wet zenith tropospheric path delay (in millimeters), dry component of zenith tropospheric path delay (in millimeters), total zenith tropospheric path delay (in millimeters), wind azimuth (in degrees) from where the wind blows, wind speed (in meters per second), rain accumulation since last measurement, and hail detected since last measurement. Also, included in the file are the met sensor type/model and its approximate geocentric position in X, Y, Z (ITRF or WGS-84) coordinates.

6.3.2 RTCM Format

While RINEX is the standard GNSS data format for receiver-independent exchange of recorded raw GNSS data for post-processing, it is not applicable for real-time data transmission in real-time positioning. RTCM is the industry standard format to enhance interoperability between different types and brands of GNSS receivers in real-time operations. The RTCM SC-104 has introduced formats and protocols that are accepted as international standards, evolving over the years (e.g., by introducing new message types [1]) with the technology, and as new satellite systems and signals emerged. The Networked Transport of RTCM via Internet Protocol (NTRIP) has been the standard protocol for delivering GNSS data via the Internet, although RT-IGS[12] was also proposed for a similar purpose [13].

The Radio Technical Commission for Maritime (RTCM) Services (rtcm.org) has various special committees that set the standards for international maritime radionavigation and radiocommunication. RTCM Special Committee (SC) 104 is the one for GNSS correction signals, and thus RTCM SC-104 format is the data format for real-time GNSS correction messages.

Different versions of RTCM SC-104 formats that have been developed over the years include, for example [33–36]:

- RTCM 2.0 (code corrections for DGPS)
- RTCM 2.1 (code+phase corrections for RTK GPS)
- RTCM 2.2 (code+phase corrections for RTK GPS+GLONASS)

[12] Real-Time IGS (RT-IGS) protocol was proposed by RT-IGS Working Group, for delivery of RT-IGS message types over the internet.

- RTCM 2.3 (code+phase corrections for RTK GPS+GLONASS+GPS antenna definition, primarily L1 only)
- RTCM 3.0 (code+phase corrections for RTK GPS+GLONASS+GPS antenna definition+Network RTK)
- RTCM 3.1 (code+phase corrections for RTK GPS+GLONASS+GPS antenna definition+Network RTK)
- RTCM 3.2 & 3.3/RTCM MSM (code+phase corrections for RTK GPS+ GLONASS+Galileo+QZSS+BDS+SBAS+Network RTK+PPP). Modernized with Multiple Signals Messages (MSM) for generic inclusion of new constellations and signals, and State Space Representation (SSR), a new message for PPP applications

RTCM SC-104 was originally set up in 1983 to develop standards for differential GPS. The first outcome was RTCM 1, which was replaced by RTCM 2.0 (in 1990) due to implementation problems. Version 2.1 added Real-Time Kinematic (RTK) messages (in 1994). Version 2.2 expanded differential operation to GLONASS (in 1998). Version 2.3 added several new messages to improve RTK (in 2001). Inefficiency of Version 2 messages (e.g. use of high bandwidth) led to the development of an improved format, RTCM 3, starting in 2004. Version 3 was primarily aimed at improving RTK and supporting network RTK. Working Groups on Network RTK, Internet Protocol, Coordinate Transformations, Reference Station Integrity Monitoring, GLONASS, Galileo, and other areas emerged since then.

Multiple Signal Messages (MSM) format has seven message types for each constellation. These include:

- MSM1 (pseudoranges for DGNSS uses)
- MSM2 (phaseranges for RTK uses)
- MSM3 (pseudoranges (code) and phaseranges (carrier) for RTK uses)
- MSM4 (pseudoranges and phaseranges plus CNR for RTK uses)
- MSM5 (pseudoranges, phasepanges, phaserangerate (Doppler) plus CNR)
- MSM6 (pseudoranges and phaseranges plus CNR, with high resolution)
- MSM7 (pseudoranges, phaseranges, phaserangerate (Doppler) plus CNR, with high resolution, for RTK uses)

Each constellation is assigned a message ID range for the seven messages (1071–1077 for GPS, 1081–1087 for GLONASS, 1091–1097 for Galileo, 1101–1107 for SBAS, 1111–1117 for QZSS, and 1121–1127 for BDS).

State Space Representation (SSR) message types include orbit and clock corrections, code biases, and URA [37]. These parameters are especially applicable for the users interested in Precise Point Positioning (PPP) filtering. The parameters are assigned message IDs for both GPS and GLONASS constellations, whereby 1057–1062 are for GPS parameters and 1063–1068 are for GLONASS parameters.

6.3.3 *BINEX Format*

BINEX ("BINary EXchange") format is a standard operational binary format developed by UNAVCO community, in collaboration with interested receiver manufacturers, to help achieve better data compression for near-real-time or real-time GNSS and other data streams (e.g., from GPS monitoring stations) [6]. It is used for operational and research purposes and supports observation and navigation messages for all GNSS constellations as well as metadata messages to encapsulate site-specific parameters.

Explanation of BINEX record structure and message types can be found in [37] (see, e.g., page 1217) and the official living document on the UNAVCO website [39] at https://binex.unavco.org/binex.html.

6.3.4 *Other Formats*

6.3.4.1 CMR Format for Real-Time Corrections

Although RTCM SC-104 format is the industry standard for real-time GNSS corrections, proprietary standards also exist, for example, the Compact Measurement Record (CMR) format developed by Trimble (in 1992). This was introduced by Trimble mainly due to inefficiency of the RTCM 2.x format which required relatively high bandwidth. In contrast, the CMR format was suitable for transmitting data at a lower baud rate and thus became widely adopted by the industry in parallel with the RTCM format. Similar to the RTCM format, the CMR format was also improved into a new format called CMR+ (introduced in 2009) to support significant changes to GNSS constellations and new signals, one of the improvements being faster initialization and better performance in difficult environments.

6.3.4.2 NMEA Format for Hardware Interfacing

The National Marine Electronics Association (NMEA) [26] developed the NMEA specifications standards which enable marine electronic equipment to send information to computers and to other equipment (i.e., enables hardware interoperability and interfacing between various equipment). GNSS receivers are designed with the hardware that meets NMEA requirements and hence can be interfaced to send positioning and pertinent information to computers and other marine electronic equipment. They are built with communication ports supporting protocols for sending (or receiving) NMEA messages.

Two NMEA standards exist, including NMEA 0183 (ASCII text format) and NMEA 2000 (for binary format). The NMEA message format gives the ability to interface GNSS receivers with NMEA devices (which includes GNSS receivers) and an interface software for electronic display and information systems. For example, it can be used to interface GNSS receivers with timekeeping equipment, radars, heading sensors, and other receivers.

Just like RINEX format makes it possible to *read* and *postprocess* recorded raw GNSS data from different receiver types without having to use a proprietary software for each receiver type, the NMEA format makes it easier to develop a software that can *interface* with any receiver instead of having to write a custom software interface for each receiver type. However, what the NMEA software interface can do with the information relayed from a receiver depends on the *NMEA sentences* (used to relay the information).

Detailed explanations of the legacy and modernized NMEA 0183 sentences can be found in [22] and [27], respectively. The various versions of both NMEA 0183 Interface Standard and NMEA 2000 Interface Standard can be accessed through the NMEA website at www.nmea.org [26].

References

1. Boriskin, A., Kozlov, D., & Zyryanov, G. (2012). The RTCM multiple signal messages: A new step in GNSS data standardization. In *Proceedings of the 25th International Technical Meeting of the Satellite Division of The Institute of Navigation (ION GNSS 2012), Nashville* (pp. 2947–2955).
2. Bruyninx, C., Legrand, J., Fabian, A., & Pottiaux E. (2019). GNSS metadata and data validation in the EUREF permanent network. *GPS Solutions, 23*, 106. https://doi.org/10.1007/s10291-019-0880-9
3. Chodota, M. L. (2003). The African reference frame project: Its current status. In: *UN/US International Workshop on the Use and Applications of Global Navigational Satellite Systems, December 8–12, Vienna, Austria.*
4. Cioce, V., Martínez, W., Mackern, M.V., Pérez, R., & de Freitas, S. (2018). SIRGAS: Reference frame in Latin America. *Coordinates, 14*(6), 6–10. www.mycoordinates.org/sirgas-reference-frame-in-latin-america/
5. Dow, J. M., Neilan, R. E., & Rizos, C. (2009). The International GNSS service in a changing landscape of global navigation satellite systems. *Journal of Geodesy, 83*, 191–198. https://doi.org/10.1007/s00190-008-0300-3
6. Estey, L., & Mencin, D. (2008). BINEX as a format for near-real time GNSS and other data streams, G43A-0663, AGU Fall Meet. 2008, San Francisco.
7. Geoscience Australia. (2019). Asia-Pacific reference frame (APREF). Retrieved 2019, from http://www.ga.gov.au/scientific-topics/positioning-navigation/geodesy/asia-pacific-reference-frame
8. Gurtner, W. (1994). RINEX: The receiver-independent exchange format. GPS WORLD (pp. 48–52).
9. Gurtner, W. (2002). RINEX: The receiver independent exchange format version 2.10. ftp://igs.org/pub/data/format/rinex210.txt
10. Gurtner, W., & Mader, G. (1990). Receiver independent exchange format version 2. CSTG GPS Bulletin, vol. 3, no. 3, Sept./Oct. 1990, National Geodetic Survey, Rockville.
11. Gurtner, W., & Estey, L. (2005). RINEX: The receiver independent exchange format version 2.11. ftp://igs.org/pub/data/format/rinex211.txt
12. Gurtner, W., Mader, G., & Arthur, D. (1989). A common exchange format for GPS data. CSTG GPS Bulletin, vol. 2, no. 3, May/June 1989, National Geodetic Survey, Rockville.

13. Heo, Y., Yan, T., Lim, S., & Rizos, C. (2009). International standard GNSS real-time data formats and protocols. In *IGNSS Symposium, Qld, Australia*.
14. Hofmann-Wellenhof, B., Lichtenegger, H., & Wasle, E. (2008). *GNSS–Global navigation satellite systems: GPS, GLONASS, Galileo & more* Berlin: Springer. ISBN 978-3-211-73012-6.
15. Hu, G., Dawson, J., Jia, M., Deo, M., Ruddick, R., & Johnston, G. (2011). Towards the densification of the international terrestrial reference frame in the Asia and Pacific region - Asia Pacific reference frame (APREF). In K. Satake & C.-H. Lo (Eds.), *Advances in geosciences* (Vol. 31). Solid Earth Science.
16. Huisman, L., Dawson, J., & Teunissen, P. J. G. (2011). The APREF project: First results and analysis. *Coordinates*, 14–18.
17. ICSM. (2014). Guideline for continuously operating reference stations, Special Publication 1, Version 2.1. Intergovernmental Committee on Surveying and Mapping (ICSM). 43pp.
18. IGS. (2018). RINEX: The receiver independent exchange format, Version 3.04. International GNSS Service (IGS), RINEX Working Group and Radio Technical Commission for Maritime Services Special Committee 104 (RTCM-SC104). ftp://igs.org/pub/data/format/rinex304.pdf
19. IGS. (2019). IGS network. www.igs.org/network
20. IGS. (2019). IGS site guidelines. International GNSS service. https://kb.igs.org/hc/en-us/sections/200409633-Site-Guidelines
21. Ihde, J., Baker, T., Bruyninx, C., Francis, O., Amalvict, M., Kenyeres, A., Mäkinen, J., Shipman, S., Simek, J., & Wilmes, H. (2005). Development of a European combined geodetic network (ECGN). *Journal of Geodynamics, 40*(4–5), 450–460.
22. Langley, R. B. (1995). NMEA 0183: A GPS receiver interface standard. GPS World.
23. NGS. (2018). Guidelines for new and existing continuously operating reference stations (CORS). National Geodetic Survey, National Oceanic and Atmospheric Administration. http://www.ngs.noaa.gov/PUBS_LIB/CORS_guidelines.pdf
24. NGS. (2019). CORS map. www.ngs.noaa.gov/CORS_Map/
25. NGS. (2019). Guidelines for establishing and operating CORS. NGS, NOAA, Silver Spring, MD 20910.
26. NMEA. (2019). National marine electronics association, nmea.org.
27. NMEA 0183. (2018). Standard for interfacing marine electronic devices, Version 4.11.
28. NRCan. (2019). GNSS reference station installation and operation best practices. NRCan, Ottawa, Canada.
29. Pestana, A. (2015). Reading RINEX 2.11 observation data files. https://doi.org/10.13140/RG.2.1.4888.4087
30. Petovello, M. (2015). Why are carrier phase ambiguities integer? InsideGNSS (pp. 36–38). www.insidegnss.com
31. RCMRD. (2016). african geodetic reference frame (AFREF). AFREF Newsletter No. 18. www.rcmrd.org/newletters?download=19:afref-newsletter-no-18
32. Rizos, C. (2008). Multi-constellation GNSS/ RNSS from the perspective of high accuracy users in Australia. *Journal of Spatial Sciences, 53*(2), 29–63.
33. RTCM. (2001). RTCM 10402.3 RTCM recommended standards for differential GNSS (Global Navigation Satellite Systems) Service, Version 2.3.
34. RTCM. (2004). RTCM recommended standards for differential GNSS (Global Navigation Satellite Systems) Service, Version 3.0
35. RTCM. (2007). RTCM standard 10403.1. Differential GNSS (Global Navigation Satellite Systems) Services – Version 3, with Amendment 2. RTCM Special Committee no. 104, Arlington, Virginia.
36. RTCM. (2016). RTCM 10403.3. Differential GNSS (Global Navigation Satellite Systems) Services - Version 3. RTCM SC 104, Arlington, Virginia.
37. Teunissen, P., & Montenbruck, O. (2017). *Springer handbook of global navigation satellite systems*. Berlin: Springer. ISBN 978-3-31942-928-1.

38. Torres, J. (2005). EUREF the infrastructure for geo-referencing in Europe. *GeoInformatics, 8*(6), 18–21.
39. UNAVCO. (2019). BINEX: binary exchange format. www.binex.unavco.org.
40. Wonnacott, R., & Nonguierma, A. (2016). The African geodetic reference frame. GIM International. www.gim-international.com

Chapter 7
GNSS Data Processing

7.1 Introduction

Figure 7.1 is an example illustration of two case scenarios for GNSS geodetic parameter estimations. In scenario A, satellite orbits and range observations are used as inputs to estimate unknown receiver position (and other parameters). In scenario B, the ground station positions (e.g., in the case of GNSS CORS) are already known from previous measurements, the satellite orbits are also known, but both can be re-estimated, or improved, from new (or continuous) satellite-receiver range observations. Most GNSS geodetic processing problems encompass both scenarios together.

Table 7.1 is an example of list of parameters that can be estimated within different solutions of GNSS data processing [17]. On the basis of scenario A, the starting point are RINEX data containing code and phase observables (i.e., range observations in Fig. 7.1) and satellite orbits (e.g., from merged IGS/CODE orbits). For scenario B, the reprocessed (or combined) solutions (e.g., the weekly and final solutions in Table 7.1) are based on normal equations derived after scenario A (1-day solutions) in which all parameters were estimated, potentially with some a priori information such as Bulletin A Earth rotation parameters. In the latter case, some parameters are pre-eliminated, while some (such as station coordinates and Earth rotation parameters) are fixed during reprocessing. Figure 7.2 is an example of the overall processing workflow upon which the entire chapter is based.

Chapter 5 (Sect. 5.3) discusses error mitigation methods, e.g., through differenced observables. For preprocessing, different strategies can be applied, such as using smoothed code observations to estimate ionosphere parameters, DCBs, and clocks, and different phase and code linear combinations can be formed based on two frequencies (such as L1 and L2) to reduce or eliminate certain effects in Eqs. (5.3) and (5.4).

C. A. Ogaja, *Introduction to GNSS Geodesy*,
https://doi.org/10.1007/978-3-030-91821-7_7

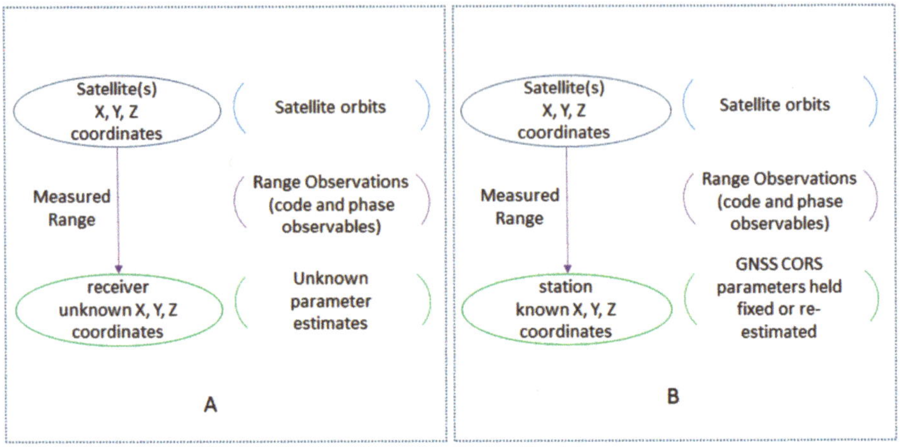

Fig. 7.1 Case scenarios of GNSS geodetic parameter estimations

Table 7.1 An example of list of parameters estimated within different data processing solutions. p indicates that the parameters are pre-eliminated, and f indicates that the parameters are fixed to the results of weekly solution [17]

	Solution		
Parameter	1-day	Weekly	Final
Station coordinates	x	x	f
Earth rotation parameters	x	x	f
Satellite orbits	x	p	x
Troposphere parameters	x	p	x
Origin of tracking network	x	x	x
Ambiguities	x	p	p
Differential code biases	x	...	...
Receiver and satellite clocks	x	...	...
Global ionosphere maps	x	...	...

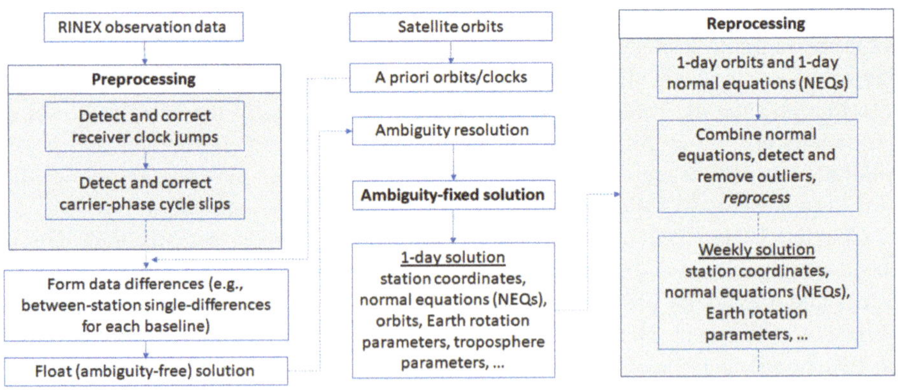

Fig. 7.2 Example of GNSS processing workflow

7.2 Preprocessing

During preprocessing, the raw GNSS data from each RINEX observation file are processed arc by arc (an observation arc is defined as a continuous satellite pass) to prepare the data before actual processing begins. The common preprocessing steps include, for example:

i. Generating smoothed code observations for receivers tracking code and phase observations on two frequencies.
ii. Checking inconsistencies between code and phase measurements, which may have been introduced by receiver "millisecond" clock jumps. If any "millisecond" clock jumps are detected, they are repaired to make the code and phase observations consistent.
iii. Detecting phase cycle slips using linear combinations. If cycle slips are detected, the linear combinations are used to determine the size of the cycle slips in each of the frequencies. The cycle slips are then either repaired or used to set up new ambiguities at the epochs of the cycle slips. In some instances, the cycle slips information can be used to connect smoothed code observations before and after the cycle slip.

7.2.1 *Smoothed Code Observations*

For code and phase observations on L_1 and L_2, phase smoothed code observations $\tilde{R}_i(t)$ where frequency index $i = 1, 2$ is given by [17]:

$$\tilde{R}_1(t) = \bar{R}_1 + \Delta\Phi_1(t) + 2\frac{f_2^2}{f_1^2 - f_2^2}(\Delta\Phi_1(t) - \Delta\Phi_2(t)) \tag{7.1}$$

$$\tilde{R}_2(t) = \bar{R}_2 + \Delta\Phi_2(t) + 2\frac{f_1^2}{f_1^2 - f_2^2}(\Delta\Phi_1(t) - \Delta\Phi_2(t)) \tag{7.2}$$

where

$$\Delta\Phi_1(t) = \Phi_1(t) - \bar{\Phi}_1 \tag{7.3}$$

$$\Delta\Phi_2(t) = \Phi_2(t) - \bar{\Phi}_2 \tag{7.4}$$

$\Phi_i(t)$ are phase measurements at epoch t, $\bar{\Phi}_i$ are mean phase measurements of the current observation arc, and $\bar{R}_i$ are mean code measurements of the current observation arc. The noise of the smoothed code observations is significantly smaller than the noise of original code observations and depends on the number of epochs in current observation arc.

Table 7.2 Types of receiver clock jumps

Type	Time tag	Pseudorange	Carrier phase
1	Jumpy	Smooth	Smooth
2	Jumpy	Jumpy	Smooth
3	Smooth	Jumpy	Smooth
4	Smooth	Jumpy	Jumpy
5	Jumpy	Smooth	Jumpy

7.2.2 Clock Jumps

Receiver clock jumps are caused by periodic resets of receiver clock in attempt to keep the receiver time synchronized with the GPS time. Since receiver clocks drift with time due to the use of low-cost internal frequency oscillators, geodetic receivers introduce periodic clock offsets when the difference between receiver time and GPS time exceeds some threshold. This causes "clock jumps" in the "observed" data, which if left unrepaired would cause incorrect positioning results. These jumps typically occur when the clock offset exceeds 1 ms in magnitude and hence are often called "millisecond" jumps. The millisecond jumps are typically jumps by an integer number of milliseconds. Table 7.2 shows the types of clock jumps that can be encountered when handling data from different receiver types (the first four are described in [10]).

When a clock jump occurs, the three quantities (time tag, pseudorange, and carrier phase measurement) in a RINEX observation file (Sect. 6.3.1 of Chap. 6) are prone to be affected by discontinuity (jump). Due to various mechanisms and reset options by different receivers and converter software, the jumps are added in different forms to these quantities.

Detecting and repairing clock jumps depends on the different jump types as shown in Table 7.2. Types 1 and 4 jumps create consistent pseudorange and phase measurements and thus can be easily handled. Types 2, 3, and 5 have inconsistencies between the code and phase measurements.

The clock jumps must be compensated when they are present in one type of raw measurement (code or phase) either by correcting for them in the affected measurement or by introducing the jumps into the smooth measurement to make both measurements consistent. However, when clock jumps are present in both code and phase measurements, they can be estimated and treated separately or lumped with estimated receiver clock error.

Clock jumps are typically detected and corrected for by comparing epoch-to-epoch differences of both the phase and code observations. Although different researchers have described different methods for handling clock jumps, for example through the use of linear combinations, methods that utilize between-epoch geometry-free and Melbourne–Wubbena linear combinations (Appendix B) would not be effective for Type 4 jumps because the clock jump effect cancels out in the derived test statistic [7].

Types 1, 2, and 5 have jumpy (uneven) time tags. Regardless of whether the pseudorange and phase measurements are consistent or not, the jumpy time tags are corrected by applying offsets corrections to make them smooth. Corresponding offsets are introduced into both the pseudorange and phase measurements. Thus, for each "smoothed" time tag, the milliseconds offset is scaled into seconds, multiplied by speed of light and applied to the pseudorange measurement. Similarly, the milliseconds offset (in seconds) is multiplied by the carrier frequency and applied to the phase measurement.

7.2.3 Cycle Slips

A cycle slip is a jump of integer number of cycles in carrier phase measurement that occurs due to instantaneous temporary loss of lock on a carrier phase signal (pseudoranges are not affected). It causes a resolved integer cycle ambiguity N to become instantly unknown again, thus affecting the correct geodetic positioning. When that happens, the correct positioning requires that either the ambiguity N be resolved again or the magnitude of the cycle slip be known exactly and corrected.

Linear combinations are used to detect and repair cycle slips (see, e.g., [4, 5, 8, 11, 17], and Appendix B for equations of different linear combinations between any two given frequencies such as L_1 and L_2).

The various linear combinations of raw measurements are formed to mitigate (i.e., reduce or remove) geometric parameters and other errors and to introduce characteristics that are beneficial to cycle slip detection (the advantages and disadvantages of the various linear combinations are included in Appendix B). For example, the linear combinations with longer wavelengths and lower noise are especially advantageous in this process.

Traditionally, linear combinations based on dual-frequency measurements are used. However, some researchers have introduced methods based on triple-frequency GNSS measurements (see, e.g., [9, 25]). Such measurements have longer wavelengths and lower measurement noise. The general form of linear combinations for triple-frequency observations is as follows:

$$\Phi_i = k_{1,i} \cdot \Phi_1 + k_{2,i} \cdot \Phi_2 + k_{3,i} \cdot \Phi_3 \tag{7.5}$$

$$R_i = k_{1,i} \cdot R_1 + k_{2,i} \cdot R_2 + k_{3,i} \cdot R_3 \tag{7.6}$$

where Φ_i and R_i are the combined phase and range observations, respectively, at epoch i; and $k_{1,i}$, $k_{2,i}$, and $k_{3,i}$ are the arbitrary coefficients of the linear combination at the epoch i. The subscripts 1, 2, and 3 represent three frequencies of any GNSS carrier signals.

The main approaches to detect and repair cycle slips include: (a) introducing additional parameters to resolve new ambiguities when a cycle slip occurs, and (b) estimating the size (or magnitude) of cycle slip and correcting for it directly at each epoch. Either of these two methods can be used in a post-processing mode. However, the first method is synonymous to re-initialization of ambiguities and may lead to long convergence time. And if there are multiple cycle slips, there will be a large number of parameters that may have to be introduced back into the model (Eq. (5.4) in Chap. 5).

For detecting the size or magnitude of a cycle slip, different methods have been used such as polynomial fitting and time differencing between epoch data. These work very well for detecting large cycle slips but may have challenges in case of small slips of 1 to 2 cycles. It is challenging to detect and repair smaller slips accurately in the presence of clock errors, atmospheric refraction, multipath noise, and so forth. This is why it is important to select carefully the linear combination to be used in the process.

7.3 Ambiguity Fixing

7.3.1 *Mathematical Model*

Consider the following system of linearized observation equations (following the background explanations in Sect. 4.2 of Chap. 4):

$$\tilde{y} = \underline{A}x + v \tag{7.7}$$

$$\underbrace{\tilde{y} - F(x_o)}_{\Delta\tilde{y}} \approx \underbrace{\left.\frac{\partial F(x)}{\partial x^T}\right|_{x_o}}_{\underline{A}} \underbrace{(x - x_o)}_{\Delta x} + \tilde{v} \tag{7.8}$$

Now let $\Delta x = [\ a;\ \ b\ ]^{\ T}$, such that the model in Eq. (7.8) can be rewritten in a new general form as (see, e.g., [18, 19, 21, 22])

$$\Delta\tilde{y} = Aa + Bb + \tilde{v} \tag{7.9}$$

where $\underline{A} = [A; B]$, a and b are the unknown parameter vectors, respectively, of orders n and p, and $\tilde{v}$ is the noise vector of the model. In the case of single differenced observations (e.g., between-station differences), the GNSS data vector $\Delta\tilde{y}$ will consist of the observed minus computed single difference (SD) phase and/or pseudorange (code) observations accumulated over all the observation epochs. The vector a are then the SD carrier phase ambiguities, expressed in units of cycles rather than range. The vector b consist of the remaining unknown parameters, such as baseline components and other parameters such as atmospheric delay parameters and hardware biases. The parameters in b are known to be real-valued (float).

For illustration, a between-station SD solution of a single baseline using multiple satellites will involve several SD ambiguity (ΔN) terms, in which case the vector a of Eq. (7.9) can be expressed as

$$a = [\ \Delta N_j]^T,\ j = 1,\ n \tag{7.10}$$

A least squares solution of Eq. (7.9) is obtained such that:

$$\min_{a,b} ||\Delta\tilde{y} - Aa - Bb||^2_{Q_{\Delta\tilde{y}\Delta\tilde{y}}} \tag{7.11}$$

For simplicity of expression in subsequent formulations, let $\Delta\tilde{y}$ be replaced with y and $Q_{\Delta\tilde{y}\Delta\tilde{y}}$ be replaced with Q_{yy} accordingly.

7.3.2 Estimation Process

As shown in Eq. (7.9), the unknown ambiguity parameters a need to be estimated together with the baseline components in b to obtain the correct positioning results. In principle, if there is no cycle slip or signal loss of lock, the ambiguity remains unchanged during the data session and hence can be set as a constant value once estimated and fixed to an integer. The procedure for solving the model (7.9) can be divided into three steps: the *float solution*, *integer ambiguity estimation*, and *fixed solution*.

7.3.2.1 Float Solution

The first step simply disregards the integer constraints on the ambiguities and applies a standard least squares adjustment, resulting in real-valued estimates of a and b, together with their variance–covariance estimates:

$$\begin{bmatrix} \hat{a} \\ \hat{b} \end{bmatrix}, \begin{bmatrix} Q_{\hat{a}} & Q_{\hat{a}\hat{b}} \\ Q_{\hat{b}\hat{a}} & Q_{\hat{b}} \end{bmatrix} \tag{7.12}$$

The solution (7.12) is commonly referred to as the *float* solution. This solution is obtained from the following system of normal equations:

$$\begin{bmatrix} A^T Q_{yy}^{-1} A & A^T Q_{yy}^{-1} B \\ B^T Q_{yy}^{-1} A & B^T Q_{yy}^{-1} B \end{bmatrix} \begin{bmatrix} \hat{a} \\ \hat{b} \end{bmatrix} = \begin{bmatrix} A^T Q_{yy}^{-1} y \\ B^T Q_{yy}^{-1} y \end{bmatrix} \tag{7.13}$$

where:

$$\hat{a} = (A^T Q_{yy}^{-1} A)^{-1} A^T Q_{yy}^{-1} y \tag{7.14}$$

$$Q_{\hat{a}} = Q_{\hat{a}\hat{a}} = (A^T Q_{yy}^{-1} A)^{-1} \tag{7.15}$$

and

$$\hat{b} = (B^T Q_{yy}^{-1} B)^{-1} B^T Q_{yy}^{-1}(y - A\hat{a}) \tag{7.16}$$

$$Q_{\hat{b}} = Q_{\hat{b}\hat{b}} = (B^T Q_{yy}^{-1} B)^{-1} \tag{7.17}$$

7.3.2.2 Integer Ambiguity Estimation

In the second step, the float estimate ($\hat{a}$) of ambiguities and its VCV are used to compute the corresponding integer estimates ($\check{a}$). This is done by mapping from the n-dimensional space of real numbers to the n-dimensional space of integers. Known examples of mapping methods include integer rounding, integer bootstrapping, and integer least squares or integer search methods ([6, pp. 214–237] of [11, 13, 18–22]).

LAMBDA method [22] is commonly used to compute the integer solution of Eq. (7.11) by solving the minimization problem:

$$\min_a (\hat{a} - a)^T Q_{\hat{a}\hat{a}}^{-1} (\hat{a} - a), \quad \text{where } a = \text{integer} \tag{7.18}$$

7.3.2.3 Fixed Solution

Finally, in the third step, the computed integer ambiguities ($\check{a}$) are subsequently applied to correct the float estimates, $\hat{b}$, of the remaining parameters. This gives the *fixed* solution as follows:

$$\check{b} = \hat{b} - Q_{\hat{b}\hat{a}} Q_{\hat{a}\hat{a}}^{-1} (\hat{a} - \check{a}) \tag{7.19}$$

where $Q_{\hat{a}\hat{a}}$ are the covariance matrices for the fixed ambiguities and $Q_{\hat{b}\hat{a}}$ are the covariance matrices between baseline components and ambiguities. It is noted here with caution that the three-step procedure may still be ambiguous depending on which mapping method is chosen. A ratio test[1] can be applied to validate the ambiguity fixing success rate [15, 20, 23]. If success rate is very close to 1, the VCV matrix of the fixed solution is obtained as

$$Q_{\check{b}\check{b}} = Q_{\hat{b}\hat{b}} - Q_{\hat{b}\hat{a}} Q_{\hat{a}\hat{a}}^{-1} Q_{\hat{a}\hat{b}} \tag{7.20}$$

[1] Acceptance test, an optional step after integer estimation, consists of deciding whether or not to accept the integer solution once integer estimates of the ambiguities have been computed.

Table 7.3 Ambiguity resolution strategies depending on the baseline length

Method	Baseline length
Melbourne–Wubbena (MW)	≤6000 km
Quasi-Ionosphere-Free (QIF)	≤2000 km
Wide-lane/narrow-lane	≤200 km
Direct solution of L1, L2 (and L5)	≤20 km

7.3.3 Baseline-Dependent Strategies

Table 7.3 shows different methods that can be used to resolve ambiguities to integers depending on the baseline length [11, 17]. The methods are based on the properties of the different linear combinations of GNSS observables (see examples in Appendix B).

The Melbourne–Wubbena (MW) method uses the linear combination of code and phase observations. Its longer wavelength increases the search space for better estimations of wide-lane ambiguities, and its geometry-free nature (i.e., not impacted by baseline length) means it is well-suited for very long baselines. It is commonly used in certain situations to first fix the wide-lane ambiguities that are then subsequently used to fix narrow-lane ambiguities that have smaller search space for integer estimations.

The Quasi-Ionosphere-Free (QIF) method solves for ambiguities without code measurements. However, since the IF linear combination still has higher order ionospheric effects that should be modeled, and the small wavelength makes ambiguity fixing difficult, it becomes necessary to also consider a priori ionosphere information such as global ionosphere models.

The wide-lane/narrow-lane method is similar to the MW method except it does not use code measurements when fixing the wide-lane ambiguities. Once the wide-lane ambiguities are fixed in the first step, they are introduced in a second step to fix the narrow-lane ambiguities. This method makes it possible to solve ambiguities for long baselines (up to hundreds of kilometers) where the L1, L2 (and L5) cannot be fixed by direct solution.

For short baselines (up to 20 km), the direct solution of the L1, L2 (and L5) is considered optimal by using the full variance–covariance information (see, e.g., in Eq. (7.12)). The atmospheric conditions are assumed to be similar for both stations (receivers), and therefore influences such as of ionospheric refraction are ignored in the single difference case.

7.4 Reprocessing

7.4.1 *Workflow*

Figure 7.3 is an example of GNSS data reprocessing using normal equations, e.g., from multiple consecutive single (hourly or daily) solutions into a single model (e.g., of multi-hour, multi-day, or weekly solutions). Similarly, reprocessing of multi-year time series data, e.g., to benefit from longer duration of measurements, additional stations added to the network, and improvements in models (for better understanding of plate tectonics and reference frames), can help determine improved coordinates and secular velocities for networks and continuously operating reference stations.

The theory of combining multiple data from normal equations is briefly covered in Chap. 4 (see also [1, 14, 24]). In the next two Sects. 7.4.2 and 7.4.3, the benefits and applications of reprocessing for epoch solutions and time series are given with examples (where available).

7.4.2 *Epoch Solutions*

For epoch solutions, some of the benefits, and applications, of reprocessing using normal equations include, for example:

1. **Combining multi-GNSS solutions** at the normal equations level, e.g., for a given baseline, process each GNSS system separately and then combine normal equations. For example, perform between-station single difference solutions for GPS and GLONASS separately and then combine their normal equations into a single solution. This circumvents the inability to resolve integer ambiguities in a GPS/GLONASS double-difference model due to the incompatible frequencies of observables. This approach also circumvents other challenges such as inter-system biases.
2. **Combining hourly solutions into a single daily solution**, and/or daily solutions into a single weekly solution, at the normal equations level. In contrast to the

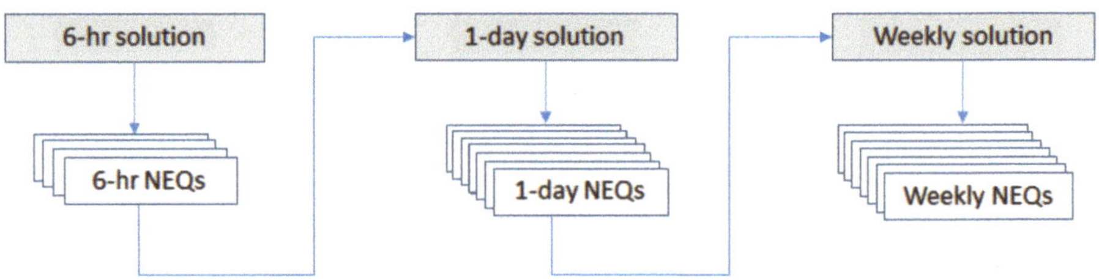

Fig. 7.3 An example of GNSS reprocessing using normal equations

multi-GNSS case above where normal equations being combined are from same epoch, the normal equations are from multiple consecutive single epoch solutions (from single- or multi-GNSS systems). For example, normal equations from multi-GNSS or GPS-only hourly solutions combines into daily solutions, and so forth.

3. **Combining solutions for a network of stations** (where, for example, in every epoch or timestamp, there are a set of *S* solutions with positions and velocities and other parameters such as polar motion, UT, and their daily rates, expressed in a particular Terrestrial Reference Frame (TRF), the sets of solutions being from different *S* entities such as IGS analysis centers) into a new combined solution where each station has positions and velocities in a combined frame at an epoch (e.g., [1]). Time series of station coordinates and TRF parameters are subsequently derived from such a combined solution (a topic of the next section).

7.4.3 Time Series

In GNSS geodesy, reprocessing is commonly applied to derive: (a) a Terrestrial Reference Frame (TRF) and (b) time series of station coordinates and their velocities. The latter plays an important role in many applications of GNSS including, for example, the study of geodynamics and crustal deformations, whereas CORS networks are the infrastructure for the realization of reference systems (i.e., deriving a terrestrial reference frame).

In a reprocessing workflow (Fig. 7.4), the normal equations covering the whole time period are accumulated to compute a TRF solution that subsequently provides a basis for the computation of station coordinates and time series of the origin of the CORS tracking network.

As part of a TRF solution, there is a 14-parameter similarity transformation between two reference frames, that is, between a new reference frame being defined and an existing older reference frame [2, 3, 12]. This usually means the existing reference frame had been defined a few years in the past, and now there are additional data and/or additional stations that need to be incorporated into the realization. Increase in data and/or station density leads to improvements in the solution.

A new solution ensures that there is a link between the new TRF and past TRF solutions, for various applications. For example, providing users with 14 transformation parameters of IGb08 with respect to IGS05 and consequently to past TRF solutions.

A subset of stable stations in an existing prior TRF solution (e.g., IGS05) are used to define the datum for the new TRF solution. The first step in data processing involves pre-elimination of some parameters. The general concept for parameter pre-elimination and the subsequent stacking of reduced normal equations for a TRF solution (per Fig. 7.4) includes:

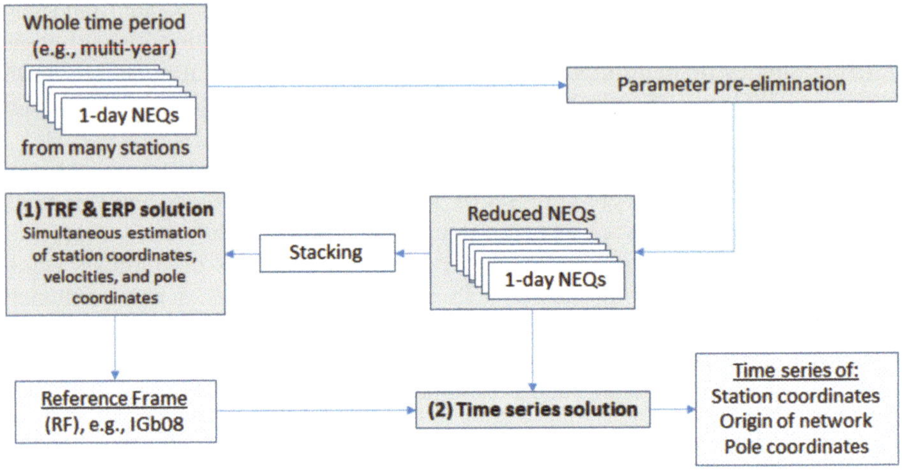

Fig. 7.4 An example of a reference frame and time series solution workflow

1. **Parameter pre-elimination**. As shown in Chap. 4, the two matrices $A^T WA$ and $A^T W\tilde{y}$ are together called the *normal equation* of a single solution and can be used in a reprocessing to regenerate results without having to repeat the steps of starting off with the raw observations. In a further illustration, let $\boldsymbol{N} = A^T WA$ and $\boldsymbol{b} = A^T W\tilde{y}$ so that:

$$\Delta\boldsymbol{x} = \boldsymbol{N}^{-1}\boldsymbol{b} \tag{7.21}$$

where the unknowns $\Delta\boldsymbol{x}$ represent an improvement for the a priori values $\boldsymbol{x_0}$ so that the updated parameters become $\boldsymbol{x} = \boldsymbol{x_0} + \Delta\boldsymbol{x}$. Parameter pre-elimination is an important operation for reducing the size of a normal equation (NEQ) by *pre-eliminating* parameters that may have been initially estimated but are no longer of interest for the reprocessing application. These may include, for example, phase ambiguities, clock parameters, and UT1 (see, e.g., [24]). Therefore, in reprocessing, these parameters are skipped (pre-eliminated from the normal equation system). For such cases the NEQ is considered to be of two parts: $\boldsymbol{x}_1$ consisting of the parameters that will be retained and $\boldsymbol{x}_2$ comprising those that will be pre-eliminated, so that:

$$\begin{bmatrix} \boldsymbol{N}_{11} & \boldsymbol{N}_{12} \\ \boldsymbol{N}_{21} & \boldsymbol{N}_{22} \end{bmatrix} \begin{bmatrix} \boldsymbol{x}_1 \\ \boldsymbol{x}_2 \end{bmatrix} = \begin{bmatrix} \boldsymbol{b}_1 \\ \boldsymbol{b}_2 \end{bmatrix} \tag{7.22}$$

By solving for $\boldsymbol{x}_2$ in second row of Eq. (7.22),

$$\boldsymbol{x}_2 = \boldsymbol{N}_{22}^{-1}(\boldsymbol{b}_2 - \boldsymbol{N}_{21} \cdot \boldsymbol{x}_1) \tag{7.23}$$

and then substituting $\boldsymbol{x}_2$ in Eq. (7.22) first row, the reduced normal equation system is now of the form:

$$\underbrace{(\boldsymbol{N}_{11} - \boldsymbol{N}_{12}\boldsymbol{N}_{22}^{-1}\boldsymbol{N}_{21})}_{\boldsymbol{N}_{(R)}} \cdot \boldsymbol{x}_1 = \underbrace{\boldsymbol{b}_1 - (\boldsymbol{N}_{12}\boldsymbol{N}_{22}^{-1} \cdot \boldsymbol{b}_2)}_{\boldsymbol{b}_{(R)}} \tag{7.24}$$

$$\boldsymbol{N}_{(R)} \cdot \boldsymbol{x}_1 = \boldsymbol{b}_{(R)} \tag{7.25}$$

The new normal equation (the matrices on the left- and right-hand sides) is smaller due to parameter pre-elimination.

2. **Stacking normal equations**. The concept of normal equation combination is illustrated in Chap. 4, Sect. 4.2.3, Eqs. (4.13)–(4.17). For a simple illustration, assume that the same set of unknown parameters $\boldsymbol{x}$ were estimated from two sets of observations, and the normal equation matrices and vectors were generated as follows:

$$\boldsymbol{N}_1 = A_1{}^T W_1 A_1\ , \quad \boldsymbol{b}_1 = A_1{}^T W_1 \tilde{y}_1 \tag{7.26}$$

$$\boldsymbol{N}_2 = A_2{}^T W_2 A_2\ , \quad \boldsymbol{b}_2 = A_2{}^T W_2 \tilde{y}_2 \tag{7.27}$$

For a combined solution of the identical parameters $\boldsymbol{x}$, the two normal equation matrices and vectors are summed up as follows:

$$(\boldsymbol{N}_1 + \boldsymbol{N}_2) \cdot \boldsymbol{x} = \boldsymbol{b}_1 + \boldsymbol{b}_2 \tag{7.28}$$

Therefore, by stacking a series of n normal equations (NEQs), one obtains the combined NEQ system [16]:

$$\tilde{\boldsymbol{N}} \cdot \tilde{\boldsymbol{x}}_c = \tilde{\boldsymbol{b}}, \text{ where, } \tilde{\boldsymbol{N}} = \sum_{i=1}^{n} A_i^T W_i A_i \text{ and } \tilde{\boldsymbol{b}} = \sum_{i=1}^{n} A_i^T W_i \tilde{y}_i \tag{7.29}$$

Stacked normal equations are used in a combined TRF and ERP solution. For further reading and examples of detailed formulations for such a combined solution, see examples in [1, 3, 24].

Once a new reference frame is obtained from the combined TRF and ERP solution, it is subsequently used for datum definition to derive the time series solution based on the reduced normal equations saved from the parameter pre-elimination step. For example, a time series of transformation parameters from each of the 1-day reduced normal equations to the TRF solution can be derived on the basis of the expression:

$$\boldsymbol{X}_{TRF} = \boldsymbol{X}_{day} + \boldsymbol{T} + D\boldsymbol{X}_{day} + \boldsymbol{R}\boldsymbol{X}_{day} + \cdots \tag{7.30}$$

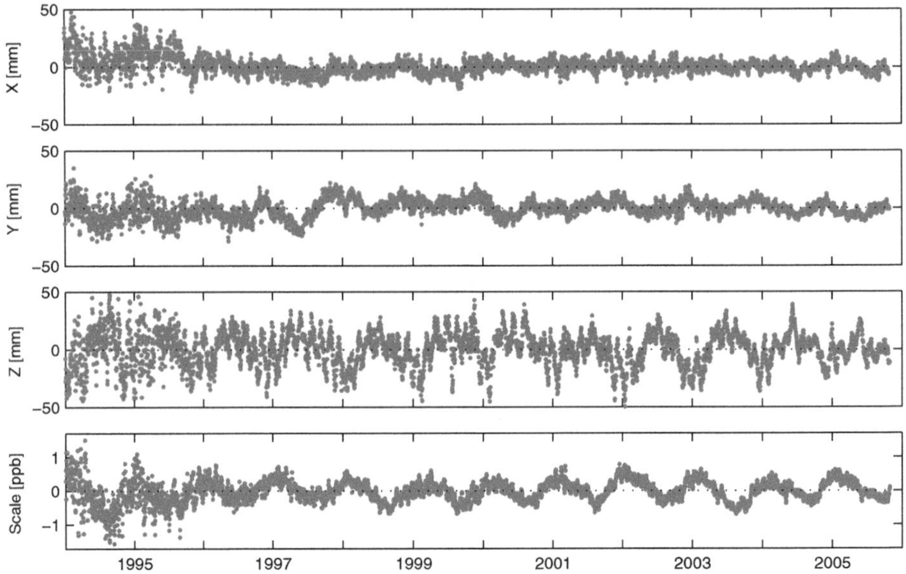

Fig. 7.5 Time series of origin (geocenter) and scale of a tracking network (From *Reprocessing of a Global GPS Network*, ISBN 978-3-7696-5052-5 [17]. Used by permission)

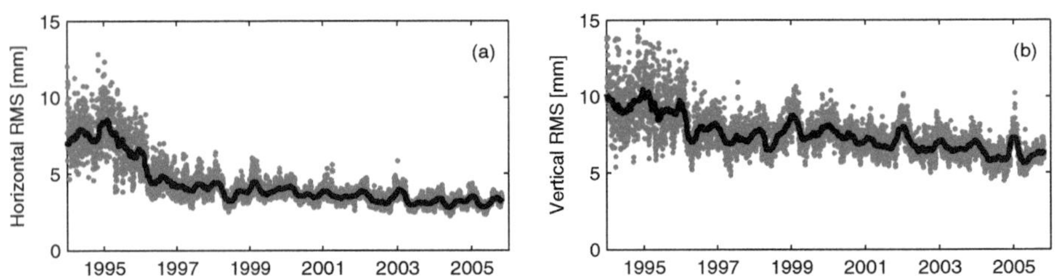

Fig. 7.6 Time series of RMS of daily 7-parameter transformations between the daily time series solution and a reference frame solution from a tracking network. The solid line indicates a 100-day median (From *Reprocessing of a Global GPS Network*, ISBN 978-3-7696-5052-5 [17]. Used by permission)

The time series of the individual solutions contributing to the combined TRF solution can then be assessed for scatter of residuals with respect to the linear model of the combined TRF solution.

Figures 7.5 and 7.6 are examples of time series evolution [17]. The origin was estimated simultaneously with station positions and ERPs (as shown earlier in Table 7.1). However, the scale time series were determined as part of a 7-parameter similarity transformation between the daily station positions and a reference frame solution based on Eq. (7.30).

The benefit of reprocessing is the fact that more stations built, more data observed, better equipment, and improved models lead to better estimations. For example, as can be seen in both Figs. 7.5 and 7.6, the first two years show larger values and scatter due to the sparse tracking network, and these get smaller with time as the number of stations increases.

References

1. Altamimi, Z., Sillard, P., & Boucher, C. (2006). CATREF software: Combination and analysis of terrestrial reference frames.
2. Altamimi, Z., & Collilieux, X. (Eds.) (2013). ITRF combination: Theoretical and practical considerations and lessons from ITRF2008. In *Reference Frames for Applications in Geosciences* (pp. 7–12). International Association of Geodesy Symposia 138, ISBN 978-3-642-32997-5.
3. Altamimi, Z., Rebischung, P., Métivier, L., & Collilieux, X. (2016). ITRF2014: A new release of the International Terrestrial Reference Frame modeling nonlinear station motions. *Journal of Geophysical Research: Solid Earth, 121*, 6109–6131. https://doi.org/10.1002/2016JB013098
4. Blewitt, G. (1990). An automatic editing algorithm for GPS data. *Geophysical Research Letters, 17*(3), 199–202.
5. Cai, C., Liu, Z., Xia, P., et al. (2013). Cycle slip detection and repair for undifferenced GPS observations under high ionospheric activity. *GPS Solutions, 17*, 247–260. https://doi.org/10.1007/s10291-012-0275-7
6. Chang, X., Yang, X., & Zhou, T. (2005). MLAMBDA: A modified LAMBDA method for integer least-squares estimation. *Journal of Geodesy, 79*, 552–565.
7. Deo, M. (2015, July 15–16). Cycle slip and clock jump repair with multi-frequency multi-constellation GNSS data for precise point positioning. In *IGNSS Symposium, Gold Coast, Australia*.
8. El-Tokhey, M. E., Sorour, T. F., Ragheb, A. E., & Moursy, M. O. (2014). GPS cycle slips detection and repair through various signal combinations. *International Journal of Modern Engineering Research (IJMER), 4*(11), 247–260.
9. Gu, X., & Zhu, B. (2017). Detection and correction of cycle slip in triple-frequency GNSS positioning. *IEEE Access, 5*, 12584–12595.
10. Guo, F., & Zhang, X. (2014). Real-time clock jump compensation for precise point positioning. *GPS Solutions 18*(1): 41–50. https://doi.org/10.1007/s10291-012-0307-3
11. Hofmann-Wellenhof, B., Lichtenegger, H., & Wasle E. (2008). *GNSS–Global Navigation Satellite Systems: GPS, GLONASS, Galileo, & more*. Vienna: Springer. ISBN 978-3-211-73012-6.
12. IERS Conventions. (2010). Terrestrial reference systems and frames. In G. Petit & B. Luzum (Eds.), *IERS Technical Note No. 36* (Chap. 4). ISBN 3-89888-989-6.
13. Li, B., Verhagen, S., & Teunissen, P. (2013). GNSS integer ambiguity estimation and evaluation: LAMBDA and Ps-LAMBDA. *Lecture Notes in Electrical Engineering, 244*, 291–301. https://doi.org/10.1007/978-3-642-37404-3-26
14. NASA. (1992). SOLVE Mathematical Formulation.
15. Odolinski, R., Teunissen, P. J. G., & Odijk, D. (2014). First combined COMPASS/BeiDou-2 and GPS positioning results in Australia. Part II: Single- and multiple-frequency single-baseline RTK positioning. *Journal of Spatial Science, 59*(1), 25–46. https://doi.org/10.1080/14498596.2013.866913
16. Ostini, L. (2012). Analysis and quality assessment of GNSS-derived parameter time series (180p.).
17. Steigenberger, P. (2009). *Reprocessing of a global GPS network*. Munchen: Deutsche Geodatische Kommission. ISBN 978-3-7696-5052-5.
18. Teunissen, P. (1996). GPS carrier phase ambiguity fixing concepts. In *GPS for geodesy* (pp. 263–335). Lecture Notes in Earth Sciences. Springer.

19. Teunissen, P. (1998). Success probability of integer GPS ambiguity rounding and bootstrapping. *Journal of Geodesy, 72*, 606–612.
20. Teunissen, P. (2000). The success rate and precision of GPS ambiguities. *Journal of Geodesy, 74*, 321–326.
21. Teunissen, P. J. G. (2001, June 5–8). GNSS ambiguity bootstrapping: Theory and application. In *Proc. of KIS2001, Banff, Canada* (pp. 246–254).
22. Teunissen, P., & Verhagen, S. (2008). GNSS ambiguity resolution: When and how to fix or not to fix? *International Association of Geodesy Symposia, 132*, 143–148. ISBN 978-3-540-74583-9.
23. Teunissen, P. J. G., & Verhagen, S. (2009). The GNSS ambiguity ratio-test revisited: A better way of using it. *Survey Review, 41*(312), 138–151.
24. Thaller, D. (2008). *Inter-technique Combination Based on Homogeneous Normal Equation Systems Including Station Coordinates, Earth Orientation and Troposphere Parameters*. Dissertation, Technische Universität München.
25. Zhao, D., Roberts, G. W., Hancock, C. M., Lau, L., & Bai, R. (2019) A triple-frequency cycle slip detection and correction method based on modified HMW combinations applied on GPS and BDS. *GPS Solutions, 23*, 22.

Correction to: Introduction to GNSS Geodesy

Clement A. Ogaja

Correction to:
A. Ogaja, C. (2022), *Introduction to GNSS Geodesy*, https://doi.org/10.1007/978-3-030-91821-7

"The original version of this book 'Augmented Reality: A GNSS Use Case.' In: *Introduction to GNSS Geodesy*. Springer, Cham, was previously published online with author name being incorrect in chapter cite as-section due to incorrect tagging in chapter XML.

A. Ogaja, C. Instead of, **Ogaja, C. A.** The book has been updated with this change."

The updated original version of this book can be found
https://doi.org/10.1007/978-3-030-91821-7

C. A. Ogaja, *Introduction to GNSS Geodesy*,
https://doi.org/10.1007/978-3-030-91821-7_8

Appendix A
GNSS Satellite Orbit Model

Satellite orbits are determined by a process of estimating the orbital state vectors (the Cartesian vectors of position and velocity of a satellite at a specific time epoch) that uniquely defines the trajectory of the orbiting satellite in space. Such estimation is based on models of the satellite's orbital motion and the forces acting upon it. With the GNSS satellites acting as beacons in space, a knowledge of their orbits and clocks enables the estimation of GNSS positions of points on or above the Earth surface from the measurements taken to the satellites. Information on satellite's orbital parameters and clocks is transmitted in the GNSS navigation message. Precise orbits (ephemeris) computed from orbit integration and improvement are also provided by some scientific organizations.

A.1 Orbital State Vector

GNSS satellites[1] are kept in their orbits by the Earth's gravitational pull. In the Solar system, planetary bodies[2] orbit other planetary bodies in a similar manner. For example, the Moon orbits the Earth and the Earth and other planets orbit the Sun. In celestial mechanics, the motion of a body relative to a parent body is known as the *two-body problem*.[3] The equation of relative motion between two bodies of

[1] GNSS satellites are *artificial* (man-made) satellites launched in space by man.

[2] Planetary bodies orbiting other planetary bodies are *natural* satellites.

[3] "Given at any time the positions and velocities of two particles of known mass moving under their mutual gravitational force calculate their positions and velocities at any other time"—[5, p. 62].

C. A. Ogaja, *Introduction to GNSS Geodesy*,
https://doi.org/10.1007/978-3-030-91821-7

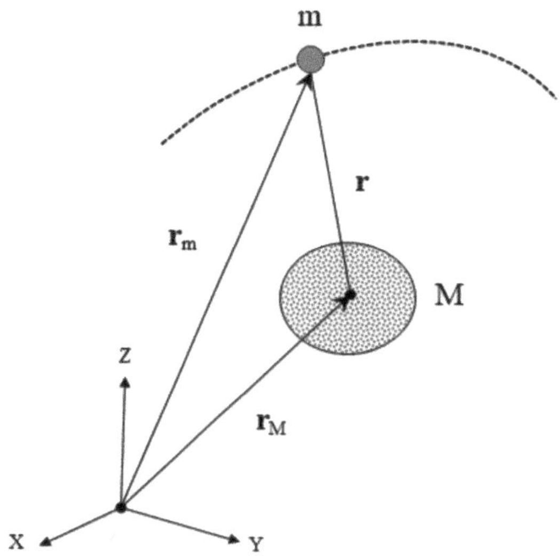

Fig. A.1 Two-body orbital motion

relative masses can be derived from *Newton's second law of motion* and his *law of universal gravitation*.[4]

Considering a system of two bodies of mass M and m as shown in Fig. A.1, their position vectors ($\mathbf{r}_m$ and $\mathbf{r}_M$, respectively) are defined in an inertial X, Y, Z cartesian coordinate system such that $\mathbf{r} = \mathbf{r}_m - \mathbf{r}_M$. Thus, $\mathbf{r}$ is their relative position vector in the defined cartesian coordinate system.[5]

According to Newtonian mechanics, the basic equation of motion of mass m relative to M, considering only the central force of attraction between them, is defined by the second-order differential equation [3–7]:

$$\ddot{\mathbf{r}} + \frac{\mu}{r^3}\mathbf{r} = 0 \tag{A.1}$$

where $\ddot{\mathbf{r}}$ denotes double differentiation w.r.t. time, $\mu = G(M + m)$, and G is the universal constant of gravitation. In this case, we are particularly interested in the motion of a GNSS satellite, an artificial Earth satellite for which the mass $m << M_E$, where M_E is the mass of the Earth. Thus, the mass of the satellite can be neglected with respect to the mass of the Earth, and therefore $\mu \approx GM_E$. The basic equation of satellite motion for a GNSS satellite is then given as

$$\ddot{\mathbf{r}} = -\frac{GM_E}{r^3}\mathbf{r} \tag{A.2}$$

where $\mathbf{r}$ is the geocentric position vector of the GNSS satellite.

[4] Derivation of the equation of motion can be found in many textbooks on celestial mechanics or satellite orbits. See, for example, [5, pp. 66–67].

[5] The cartesian coordinate system is defined with an origin in space but can also be defined with an origin located at the Earth's geocenter.

A.2 Keplerian Elements

Equation (A.2) is the vector form of a second-order differential equation from which the single and double integrations lead to [5]:

$$\dot{\mathbf{r}}(t) = \dot{\mathbf{r}}(t; a_1, ..., a_6) \tag{A.3}$$

$$\mathbf{r}(t) = \mathbf{r}(t; a_1, ..., a_6) \tag{A.4}$$

where $a_1, ..., a_6$ are the free selectable integration constants. Equation (A.4) represents the Keplerian satellite motion around the Earth, where the integration to obtain the satellite's position vector in a cartesian coordinate system applies six independent Keplerian orbital parameters $(a, e, i, \Omega, \omega, \nu)$ as illustrated in Fig. A.2. Thus, the satellite position vector at any time t is obtained from the six Keplerian parameters:

$$\mathbf{r}(t) = \mathbf{r}(t; a, e, i, \Omega, \omega, \nu) \tag{A.5}$$

The detailed steps of obtaining the solution in Eq. (A.5) from (A.2) can be found in companion textbooks (see, e.g., [5, pp. 69–74]). The basic equation of motion (Eq. (A.2)) is derived under the assumptions that only gravitational forces between the two bodies are present, that the mass of the satellite can be neglected, and that

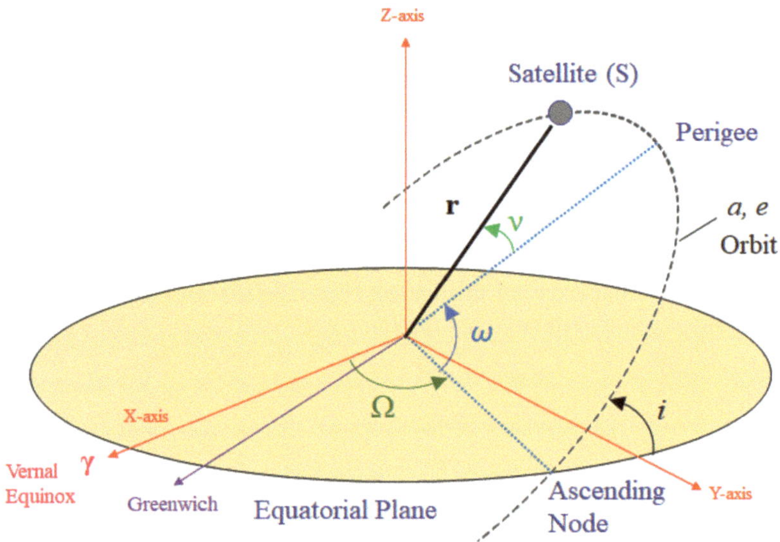

Fig. A.2 The six Keplerian orbital parameters, including: a (orbit semi-major axis), e (orbit eccentricity), i (orbit inclination), Ω (right ascension of ascending node), ω (argument of perigee), and ν (true anomaly)

the mass of the parent body (the Earth) is treated as a uniform massive body. This is an oversimplication, especially given that the Earth is an inhomogeneous mass that affects the motion of near-Earth satellites. Therefore, the Keplerian orbit (Eq. (A.5)) can only be used as a first approximation of the true satellite orbit. The effects of other forces on the orbit are examined in the next section.

A.3 Orbit Perturbations

In Eq. (A.2), the Keplerian motion assumes a GNSS satellite of negligible mass under gravitative force of a single massive body M_E. In reality, this is only a first approximation of the satellite motion in the two-body problem. Additional terms of accelerations (called perturbations) affecting the satellite motion must be added to Eq. (A.2) due to the following:

1. Earth's *size and shape, and the non-homogeneous mass distribution of the material within it*. The Earth is an ellipsoidal figure with an equatorial radius that is larger than polar radius by about 20 km, and its non-uniform density means gravitational and tidal forces depend on latitude and longitude as well as the radial distance.
2. In the complex universe, *there are other celestial bodies including the Sun, the Moon, and other planets*. This causes additional direct gravitational accelerations on the satellite motion and other indirect perturbations such as tidal forces that deform the shape of the Earth.
3. The Sun, which is the largest planetary body in the solar system, produces a non-gravitational perturbation called *solar radiation pressure*, which in turn affects the other orbit perturbations. The solar radiation pressure is zero when the satellite is in the Earth's shadow.

The perturbations are combined into a resulting vector $\mathbf{k}_s$, so that the extended equation of satellite motion becomes

$$\ddot{\mathbf{r}} = -\frac{GM_E}{r^3}\mathbf{r} + \mathbf{k}_s \tag{A.6}$$

Figure A.3 shows both the gravitational and non-gravitational accelerations (perturbations) on the satellite, from which the resulting vector $\mathbf{k}_s$ is deduced according to the following equation:

$$\mathbf{k}_s = \ddot{\mathbf{r}}_E + \ddot{\mathbf{r}}_S + \ddot{\mathbf{r}}_M + \ddot{\mathbf{r}}_e + \ddot{\mathbf{r}}_o + \ddot{\mathbf{r}}_D + \ddot{\mathbf{r}}_{SP} + \ddot{\mathbf{r}}_A \tag{A.7}$$

The gravitational perturbing accelerations include:

1. $\ddot{\mathbf{r}}_E$: the accelerations due to the non-spherical nature and inhomogeneous mass distribution of the Earth

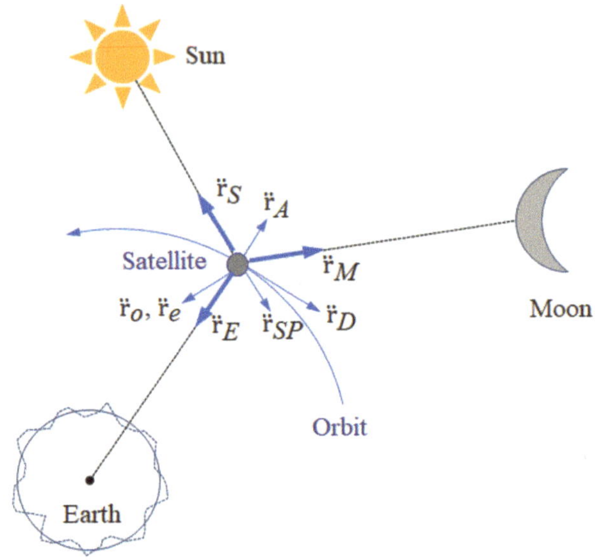

Fig. A.3 Satellite orbit perturbations

2. $\ddot{\mathbf{r}}_S, \ddot{\mathbf{r}}_M$: the accelerations due to other celestial bodies and planets, but mainly the Sun and the Moon
3. $\ddot{\mathbf{r}}_e, \ddot{\mathbf{r}}_o$: the accelerations due to the Earth and Ocean tides

The non-gravitational perturbing accelerations include:

1. $\ddot{\mathbf{r}}_D$: the accelerations due to atmospheric drag
2. $\ddot{\mathbf{r}}_{SP}, \ddot{\mathbf{r}}_A$: the accelerations due to direct and Earth-reflected solar radiation pressure

***J_2* Perturbations** The gravitational perturbations $\ddot{\mathbf{r}}_E$ due to the Earth's geoidal shape and mass distribution are modelled from the spherical harmonic expansion ([1]: Eq. 3.26) representing the Earth's potential:

$$\begin{aligned} V = \frac{\nu}{r}\Bigg[1 - \sum_{n=2}^{\infty}\left(\frac{a_E}{r}\right)^n J_n P_n(\sin\phi) \\ + \sum_{n=2}^{\infty}\sum_{m=1}^{n}\left(\frac{a_E}{r}\right)^n (C_{nm}\cos m\lambda + S_{nm}\sin m\lambda) P_{nm}(\sin\phi)\Bigg] \end{aligned} \tag{A.8}$$

where a_E is the Earth's equatorial radius, $r = |\mathbf{r}|$ is the satellite's geocentric distance (see, Figs. A.1 and A.2), ϕ and λ are the satellite's geocentric latitude and longitude. $J_n = -C_{n0}, C_{nm}, S_{nm}$ are zonal ($m = 0$) and tesseral ($m \neq 0$) coefficients; P_n and P_{nm} are the Legendre polynomials.

The even-degree zonal harmonics (coefficients) of the geopotential (the second, fourth, sixth, and so on) produce secular perturbations in ω, Ω, and M, and short-period perturbations in all six Keplerian elements, while odd-degree zonal harmonics (coefficients) produce long-period perturbations in all elements except the semi-major axis (see, for example, [5, p. 94]).

The Earth's flattening and equatorial bulge causes the largest departure of the Earth's shape from a spherical form. This departure can be represented by the second zonal harmonic (i.e., coefficient of the second order and degree zero in Eq. (A.8)), J_2 ($= C_{20}$), which models the Earth's ellipsoidal shape with uniform density. The J_2 (oblateness) perturbation causes a rotation of perigee (rotation of the major axis in the orbital plane), nodal variations (rotation of the satellite's orbit in the equatorial plane), and secular perturbations of the mean anomaly M. These are formulated as follows [3, 5]:

$$\dot{a} = \dot{e} = \dot{i} = 0 \tag{A.9}$$

$$\dot{\omega} = J_2 \frac{3n_0}{4(1-e^2)^2} \left[\frac{a_E}{a}\right]^2 (1 - 5\cos^2 i) \tag{A.10}$$

$$\dot{\Omega} = J_2 \frac{3n_0}{2(1-e^2)^2} \left[\frac{a_E}{a}\right]^2 \cos i \tag{A.11}$$

$$\dot{M} = n_0 - J_2 \frac{3n_0}{4(1-e^2)^{3/2}} \left[\frac{a_E}{a}\right]^2 (3\cos^2 i - 1) \tag{A.12}$$

where n_0 ($= 2\pi/T$) is the mean motion for the two-body problem. Thus, J_2 (the second harmonic of the geopotential)[6] does not produce secular perturbations in the orbital elements a, e, i. The J_2 perturbations in ω, Ω, and M are all dependent upon the inclination of the orbit.

From Eq. (A.10), it can be shown that the variation of ω vanishes for $i \approx 63.4°$ or $116.6°$, and from Eq. (A.11) it can be shown that the satellites in polar orbits ($i = 90°$) do not experience nodal variations ($\dot{\Omega}$) caused by J_2 (cf. the effect is maximum for equatorial orbits).

Perturbations by the Sun and the Moon $\ddot{\mathbf{r}}_S$ and $\ddot{\mathbf{r}}_M$, the perturbing accelerations caused by the gravitational attractions of the Sun and the Moon on the satellite, are given by

[6] J_2 is oblateness of the Earth, commonly called second harmonic of the geopotential.

$$\ddot{\mathbf{r}}_S = Gm_S\left(\frac{\mathbf{r}_S - \mathbf{r}}{|\mathbf{r}_S - \mathbf{r}|^3} - \frac{\mathbf{r}_S}{\mathbf{r}_S^3}\right)$$
$$\ddot{\mathbf{r}}_M = Gm_M\left(\frac{\mathbf{r}_M - \mathbf{r}}{|\mathbf{r}_M - \mathbf{r}|^3} - \frac{\mathbf{r}_M}{\mathbf{r}_M^3}\right) \tag{A.13}$$

where $\mathbf{r}$ is the satellite's geocentric position vector in the defined cartesian coordinate system, $\mathbf{r}_S$ is the Sun's geocentric position vector, $\mathbf{r}_M$ is the Moon's geocentric position vector, $Gm_S \approx 1325\,.\,10^8\,\mathrm{km}^3\,\mathrm{s}^{-2}$ is the constant for the mass of the Sun, and $Gm_M \approx 49\,.\,10^2\,\mathrm{km}^3\,\mathrm{s}^{-2}$ is the constant for the mass of the Moon. Cartesian coordinates (orbital ephemerides) of the Sun, the Moon, and other planets are available from the Jet Propulsion Laboratory (JPL) and other sources such as the Astronomical Almanac.

The acceleration on a GPS satellite is about $2\,.\,10^{-6}\,\mathrm{m/s}^2$ for the Sun and $5\,.\,10^{-6}\,\mathrm{m/s}^2$ for the Moon [5]. The influence of other planets, considered negligible, is about $3\,.\,10^{-10}\,\mathrm{m/s}^2$ [5, p. 101].

Perturbations due to Solid Earth Tides and Ocean Tides $\ddot{\mathbf{r}}_e$ and $\ddot{\mathbf{r}}_o$, the accelerations due to the Earth and Ocean tides, are considered as *indirect gravitational effects* of the Sun and the Moon on the satellite. This is because the tidal effects of the Sun and the Moon on Solid Earth and oceans change the Earth's gravitational potential, which in turn causes these two additional accelerations acting on the satellite.

The acceleration of the satellite caused by Solid Earth tides is given by the formula [5, p. 101]:

$$\ddot{\mathbf{r}}_e = \frac{k_2}{2}\frac{Gm_d}{r_d^3}\frac{a_e^5}{r^4}(3 - 15\cos^2\theta)\frac{\mathbf{r}}{r} + 6\cos\theta\frac{\mathbf{r}_d}{r_d} \tag{A.14}$$

where m_d is the mass of the disturbing body (Sun, Moon); $\mathbf{r}_d$ is the geocentric position vector of the disturbing body; θ is the angle between the geocentric position vector $\mathbf{r}$ of the satellite and $\mathbf{r}_d$; and k_2 is known as Love number (describes elasticity of the Earth's body).

The indirect effect due to ocean tides, $\ddot{\mathbf{r}}_o$, is not easy to model and is in the order of $10^{-9}\,\mathrm{m/s}^2$ [1]. Thus, the effect of ocean tides on satellite orbits is considered very small,[7] with periods of between ∼10 days and ∼100 days, the largest influence being on i and Ω orbital elements. A global tide model with coefficients can be applied to compute for each point on the ocean surface, tidal heights, and tidal-induced mass variations [5].

[7] "For GPS satellites the acceleration is of the order of $5 \cdot 10^{-10}\,\mathrm{m/s}^2$ (corresponding to less than 1 m after 2 days)."

Detailed formulas for the computation of Solid Earth and ocean tides can be found in the IERS Technical Notes (e.g., [2]). Unmodeled perturbations due to Solid Earth tides and ocean tides cause the geocentric position of an observing site to vary with time. Such variations fall under receiver-dependent biases and can be taken into account in the observation equation.

Perturbations due to Atmospheric Drag $\ddot{\mathbf{r}}_D$, the acceleration caused by atmospheric drag due to the interaction between the satellite and particles of the atmosphere, affects low-Earth-orbit (LEO) satellites but **has no effect on GNSS satellites**[8] that are in higher orbits. Thus, for GNSS satellites, the term $\ddot{\mathbf{r}}_D$ in Eq. (A.7) may be neglected. The effect on a LEO satellite depends on factors such as geometry, velocity, and orientation of the satellite as well as the density, temperature, and the composition of atmospheric gases. The perturbation is in the direction opposite to the force of the atmospheric resistance and is given by [5]

$$\ddot{\mathbf{r}}_D = -\frac{1}{2} C_D \rho(\mathbf{r}, t) \frac{A}{m_s} (\dot{\mathbf{r}} - \dot{\mathbf{r}}_a) |\dot{\mathbf{r}} - \dot{\mathbf{r}}_a| \tag{A.15}$$

where m_s is the mass of satellite, A is the effective cross-sectional area of the satellite, C_D is the drag coefficient (satellite specific), $\rho(\mathbf{r}, t)$ is the density of the atmosphere near the satellite, $\mathbf{r}, \dot{\mathbf{r}}$ are the position and velocity vectors of the satellite, and $\dot{\mathbf{r}}_a$ is the velocity of the atmosphere near the satellite.

With the assumption that the atmosphere rotates rigidly with the Earth, the relative velocity of a LEO satellite, orbiting at an altitude of 2000 km or less, with respect to the atmosphere, can be obtained as

$$\dot{\mathbf{r}} - \dot{\mathbf{r}}_a = \begin{pmatrix} \dot{x} + \dot{\theta} y \\ \dot{y} - \dot{\theta} x \\ \dot{z} \end{pmatrix} \tag{A.16}$$

where x, y, z are the satellite coordinates in a geocentric equatorial coordinate system, and $\dot{\theta}$ is the Earth rotation rate.

Perturbations due to Solar Radiation Pressure Solar radiation pressure is also considered to have negligible influence on GNSS satellites. The perturbation has two effects on a satellite, $\ddot{\mathbf{r}}_{SP}$ and $\ddot{\mathbf{r}}_A$, the accelerations due to direct and Earth-reflected solar radiation pressure. The latter effect (Earth-reflected) is also commonly referred to as *albedo*.

The direct perturbation, $\ddot{\mathbf{r}}_{SP}$, is based on factors such as satellite's surface reflectivity and surface area, solar flux, and the distance between the satellite and the Sun. It can be modeled using

---

[8] GNSS satellites are medium-Earth-orbit (MEO) satellites at altitudes of ≈20,000 km (much higher than LEO satellites). The effect of atmospheric drag decreases rapidly with increasing altitude and has no effect for GNSS satellites.

$$\ddot{\mathbf{r}}_{SP} = \upsilon P_S \frac{C_r O}{m} (\mathrm{AU})^2 \frac{(\mathbf{r} - \mathbf{r}_S)}{|\mathbf{r} - \mathbf{r}_S|^3} \tag{A.17}$$

where (AU) is the Astronomical Unit ($1.5 \cdot 10^8$ km), P_S is a constant (the quotient of solar flux and velocity of light in the Astronomical Unit), C_r is the factor of reflectivity for the satellite surface (1.95 for aluminum), O is the cross-sectional area of the satellite as seen from the Sun, m is the satellite's mass, $\mathbf{r}$ and $\mathbf{r}_S$ are the geocentric position vectors of the satellite and of the Sun in the space-fixed equatorial system, and υ is the shadow function ($\upsilon = 0$ when satellite is in the Earth's shadow; $\upsilon = 1$ when satellite is in the sunlight; and $0 < \upsilon < 1$ when satellite is in half shadow). The perturbation is extremely difficult to model as the term varies unpredictably over the year, and other factors such as reflective properties and the irregular shape of the satellites make it hard to determine area-to-mass ratio (O/m).

The albedo part of the radiation pressure is also difficult to model due to the variations in the distribution of land, sea, and clouds, but in most cases it is less than 10% of the direct radiation. The estimate for MEO satellites is very small (lies between 1 and 2%) and can be neglected for GNSS satellite orbit computations, except for very long orbital arcs.

References

1. Hofmann-Wellenhof, B., Lichtenegger, H., & Wasle, E. (2008). *GNSS: Global navigation satellite systems*. Vienna: Springer.
2. IERS Conventions. (2010). In G. Petit & B. Luzum (Eds.), *IERS Technical Note 36* (179 pp.). Frankfurt am Main: Verlag des Bundesamts für Kartographie und Geodäsie. ISBN:3-89888-989-6.
3. Kaula, W. M. (2000). *Theory of satellite geodesy: Applications of satellites to geodesy*. New York: Dover Publications.
4. Montenbruck, O., & Gill, E. (2000). *Satellite orbits: Models, methods and applications*. Berlin, Heidelberg: Springer.
5. Seeber, G. (2003). *Satellite geodesy* (2nd ed.). New York: Walter de Gruyter. ISBN:3-11-017549-5.
6. Subirana, J. S., Zornoza, J. M., & Hernández-Pajares, M. (2013). *GNSS data processing. Volume I: Fundamentals and algorithms*. ESA TM-23/1.
7. Vallado, D., & Wertz, J. (2013). *Fundamentals of astrodynamics and applications* (4th ed.). Cleveland: Microcosm Press.

Appendix B
GNSS Linear Combinations

This appendix looks at the different forms of linear combinations using GNSS observables. The concise presentation considers the general forms from which different linear combinations are derived on the basis of the combination coefficients, or factors, for each of the observables when combining them. The appendix also summarizes the advantages and disadvantages of each of the linear combinations for considerations when using to solve GNSS data processing problems.

B.1 Dual-Frequency Model

B.1.1 General Form

Given observables at two frequencies, the phase and code linear combinations can be expressed in the more general form as [3, 5]

$$\Phi_i = k_{1,i} \cdot \Phi_1 + k_{2,i} \cdot \Phi_2 \tag{B.1}$$

$$R_i = k_{1,i} \cdot R_1 + k_{2,i} \cdot R_2 \tag{B.2}$$

where $k_{1,i}$ and $k_{2,i}$ are the linear combination coefficients, and the noise of the linear combination, at epoch i, is given by

$$\sigma_i = \sigma_0 \cdot \sqrt{k_{1,i}^2 + k_{2,i}^2} \tag{B.3}$$

assuming σ_0 is the observations noise for both frequencies.

C. A. Ogaja, *Introduction to GNSS Geodesy*,
https://doi.org/10.1007/978-3-030-91821-7

The phase linear combination Φ_i has the phase ambiguity term that can be expressed in the general form of

$$\lambda_i \cdot N_i = k_{1,i} \cdot \lambda_1 \cdot N_1 + k_{2,i} \cdot \lambda_2 \cdot N_2 \tag{B.4}$$

where λ_i is the wavelength of the linear combination, and the integer*ness* of the ambiguity term N_i depends on the coefficients $k_{1,i}$ and $k_{2,i}$.

B.1.2 Ionosphere-Free

$$\Phi_{IF} = \frac{1}{f_1^2 - f_2^2}(f_1^2\Phi_1 - f_2^2\Phi_2) \tag{B.5}$$

$$R_{IF} = \frac{1}{f_1^2 - f_2^2}(f_1^2 R_1 - f_2^2 R_2) \tag{B.6}$$

For between-station single differenced observables:

$$\Delta\Phi_{IF} = \frac{1}{f_1^2 - f_2^2}(f_1^2\Delta\Phi_1 - f_2^2\Delta\Phi_2) \tag{B.7}$$

$$\Delta R_{IF} = \frac{1}{f_1^2 - f_2^2}(f_1^2\Delta R_1 - f_2^2\Delta R_2) \tag{B.8}$$

Advantages The first-order ionospheric effects are eliminated.

Disadvantages The noise level increases by a factor of three compared to the original L_1 and L_2 observables. The smaller wavelength λ_{IF} (0.6 cm) makes it impossible to resolve ambiguity (N_{IF} in (B.5), ΔN_{IF} in (B.7)). However, to circumvent this problem, if the wide-lane ambiguity N_{WL} is known, the ambiguity resolution can be made possible by, for example, replacing the L_2 ambiguity (N_2 in Eq. (B.4)) by $N_2 = N_1 - N_{WL}$ so that

$$\lambda_{IF} \cdot N_{IF} = \frac{c}{f_1 + f_2} \cdot N_1 + \frac{c \cdot f_2}{f_1^2 - f_2^2} \cdot N_{WL} \tag{B.9}$$

which becomes a narrow-lane ambiguity ($\lambda_{IF} \approx 10.7$ cm). And subsequently, Eq. (B.5) modifies into

$$\Phi_{IF} = \frac{1}{f_1^2 - f_2^2}(f_1^2\Phi_1 - f_2^2(\Phi_2 + \lambda_2 N_{WL})) \tag{B.10}$$

B.1.3 Geometry-Free

$$\Phi_{GF} = \Phi_1 - \Phi_2 \tag{B.11}$$

$$R_{GF} = R_1 - R_2 \tag{B.12}$$

Advantages Receiver clock error and geometry, i.e., the geometric range (satellite orbits and station coordinates), are eliminated. It is suitable for the estimation of ionosphere parameters.

Disadvantages Ambiguities are still included, but the ambiguity term is not integer anymore.

B.1.4 Wide-Lane

$$\Phi_{WL} = \frac{1}{f_1 - f_2}(f_1\Phi_1 - f_2\Phi_2) \tag{B.13}$$

$$R_{WL} = \frac{1}{f_1 - f_2}(f_1 R_1 - f_2 R_2) \tag{B.14}$$

For between-station single differenced observables:

$$\Delta\Phi_{WL} = \frac{1}{f_1 - f_2}(f_1\Delta\Phi_1 - f_2\Delta\Phi_2) \tag{B.15}$$

$$\Delta R_{WL} = \frac{1}{f_1 - f_2}(f_1\Delta R_1 - f_2\Delta R_2) \tag{B.16}$$

Advantages The resulting longer wavelength ($\lambda_{WL} \approx 86.2$ cm for GPS L1 and L2) is useful for ambiguity resolution and cycle slips detection.

Disadvantages Noise is greater compared to the original signals.

B.1.5 Melbourne–Wubbena

Here, both code and phase measurements are used together. It is effectively the difference between the wide-lane phase and narrow-lane code.

$$\Phi_{MW} = \frac{1}{f_1 - f_2}(f_1\Phi_1 - f_2\Phi_2) - \frac{1}{f_1 + f_2}(f_1 R_1 + f_2 R_2) \tag{B.17}$$

Advantages It has wide-lane wavelength ($\lambda_{WL} \approx 86.2\,\text{cm}$ for GPS L1 and L2), hence useful for wide-lane ambiguity resolution. It is geometry-free, hence not impacted by baseline length [6]. The effects of ionosphere, troposphere, and clocks are eliminated.

Disadvantages Higher noise due to the use of code signals [6].

B.2 Triple-Frequency Model

B.2.1 General Form

The general form of linear combinations for triple-frequency observations is as follows (see, e.g., [2, 7]):

$$\Phi_i = k_{1,i} \cdot \Phi_1 + k_{2,i} \cdot \Phi_2 + k_{3,i} \cdot \Phi_3 \tag{B.18}$$

$$R_i = k_{1,i} \cdot R_1 + k_{2,i} \cdot R_2 + k_{3,i} \cdot R_3 \tag{B.19}$$

where Φ_i and R_i are the combined phase and range observations, respectively, at epoch i; and $k_{1,i}, k_{2,i}$, and $k_{3,i}$ are the arbitrary coefficients of the linear combination at the epoch i. The subscripts 1, 2, and 3 represent three frequencies of any GNSS carrier signals.

B.2.2 Code-Phase Model

A code-phase linear combination from code and carrier phase measurements of triple-frequency signals [1] can be expressed as

$$\Phi_{tf} = (k_1 \cdot R_1 + k_2 \cdot R_2 + k_3 \cdot R_3) + (l_1 \cdot \Phi_1 + l_2 \cdot \Phi_2 + l_3 \cdot \Phi_3) \tag{B.20}$$

where the subscripts of the measurements denote the frequency, and k_i and l_i ($i = 1, 2, 3$) are the corresponding coefficients.

Using observation equations discussed in Chap. 5, the code-phase linear combination observation equation can be formed as follows:

$$\begin{aligned}\Phi_{tf} &= (k_1 + k_2 + k_3 + l_1 + l_2 + l_3)G \\ &+ c(B_{(k_1,k_2,k_3)} - b_{(k_1,k_2,k_3)} + B_{(l_1,l_2,l_3)} - b_{(l_1,l_2,l_3)}) \\ &- (l_1\lambda_1 N_1 + l_2\lambda_2 N_2 + l_3\lambda_3 N_3) \\ &+ \beta_{k_1-l_1,k_2-l_2,k_3-l_3} d_{ion(1)} + \upsilon_{tf}\end{aligned} \tag{B.21}$$

where G is the geometric part expressed as

$$\begin{aligned}G &= |(\boldsymbol{P} + \boldsymbol{E} + \boldsymbol{O}) - (\boldsymbol{p} + \boldsymbol{e} + \boldsymbol{o})| \\ &+ c\delta t_{rcv} - c\delta t^{sat} + d_{trop} + d_{orb}\end{aligned} \tag{B.22}$$

$B_{(k_1,k_2,k_3)}$ and $B_{(l_1,l_2,l_3)}$ are the receiver's code and phase hardware biases, respectively, and $b_{(k_1,k_2,k_3)}$ and $b_{(l_1,l_2,l_3)}$ are the satellite's code and phase hardware biases, respectively. $d_{ion(1)}$ is the ionospheric delay of L_1, and β is defined according to the following formulation [4]:

$$\beta_{k_1,k_2,k_3} = {f_1}^2 \left(\frac{k_1}{{f_1}^2} + \frac{k_2}{{f_2}^2} + \frac{k_3}{{f_3}^2} \right) \tag{B.23}$$

Generally, Eq. (B.21) becomes geometry-free (GF) if $k_1+k_2+k_3+l_1+l_2+l_3 = 0$ and geometry-preserving if $k_1 + k_2 + k_3 + l_1 + l_2 + l_3 = 1$.

B.3 Important Factors

From the various possible number of linear combinations, only the ones that fulfill some important criteria for the combined signals are of interest. For example, it is important to consider *reasonably long(er) wavelengths to help ambiguity fixing*, *the integer coefficients to produce integer ambiguities*, *low ionospheric influence*, and *the resultant observation noise*.

B.3.1 Wavelengths

Table B.1 shows the advantages of wide lane in giving longer wavelengths.

B.3.2 Coefficients

See Table B.2.

Table B.1 Wavelengths of wide-lane and narrow-lane linear combinations of carrier phases

System	Signals	Wide-lane wavelength	Narrow-lane wavelength
GPS	L1, L2	86.2 cm	10.7 cm
	L1, L5	75.1 cm	10.9 cm
	L2, L5	586.1 cm	12.5 cm
Galileo	E1, E5a	75.1 cm	10.9 cm
	E1, E5b	81.4 cm	10.8 cm
	E5b, E5a	976.8 cm	12.6 cm
GLONASS ($k = 0$)	G1, G2	84.2 cm	10.5 cm
	G1, G3	76.0 cm	10.7 cm
	G2, G3	782.7 cm	12.2 cm
BeiDou	B1, B3	102.4 cm	10.5 cm
	B1, B2	84.6 cm	10.8 cm
	B3, B2	488.4 cm	12.1 cm

Table B.2 Examples of ionosphere-free linear combinations for ambiguity resolution

System	Signals	Linear combination	Wavelength	Ambiguities	Noise factor
GPS	L1, L2	$2.5457\ \Phi_1 - 1.5457\ \Phi_2$	0.63 cm	$77\ a_1 - 60\ a_2$	2.98
	L2, L5	$12.2553\ \Phi_2 - 11.2553\ \Phi_3$	12.47 cm	$24\ a_2 - 23\ a_3$	16.64
	L1, L5	$2.2606\ \Phi_1 - 1.2606\ \Phi_3$	0.28 cm	$154\ a_1 - 115\ a_3$	2.59
Galileo	E1, E5b	$2.3932\ \Phi_1 - 1.3932\ \Phi_2$	0.15 cm	$308\ a_1 - 235\ a_2$	2.77
	E5b, E5a	$23.7527\ \Phi_2 - 22.7527\ \Phi_3$	12.60 cm	$47\ a_2 - 46\ a_3$	32.89
	E1, E5a	$2.2606\ \Phi_1 - 1.2606\ \Phi_3$	0.28 cm	$154\ a_1 - 115\ a_3$	2.59

References

1. Deng, C., et al. (2018). Triple-frequency code-phase combination determination: A comparison with the Hatch-Melbourne-Wübbena combination using BDS signals. *Remote Sensing, 10*(2), 353.
2. Gu, X., & Zhu, B. (2017). Detection and correction of cycle slip in triple-frequency GNSS positioning. *IEEE Access, 5*, 12584–12595.
3. Hofmann-Wellenhof, B., Lichtenegger, H., & Wasle, E. (2008). *GNSS–global navigation satellite systems: GPS, GLONASS, Galileo & more*. Berlin: Springer. ISBN:978-3-211-73012-6.
4. Li, B., Feng, Y., & Shen, Y. (2010). Three carrier ambiguity resolution: Distance-independent performance demonstrated using semi-generated triple frequency GPS signals. *GPS Solution, 14*, 177–184.
5. Steigenberger, P. (2009). *Reprocessing of a global GPS network*. Munchen: Deutsche Geodatische Kommission. ISBN:978-3-7696-5052-5.
6. Stressler, B., Bilich, A., Ogaja, C., & Heck, J. (2021). *Multi-GNSS single-difference baseline processing at NGS with newly developed M-PAGES software*. EGU General Assembly 2021, EGU21-5556.
7. Zhao, D., Roberts, G. W., Hancock, C. M., Lau, L., & Bai, R. (2019). A triple-frequency cycle slip detection and correction method based on modified HMW combinations applied on GPS and BDS. *GPS Solutions, 23*, 22.

Appendix C
GNSS Applications in Geohazard, Infrastructure, and Environmental Monitoring

There are many applications of GNSS geodesy, and more are being discovered through demand, research, and innovation. Both recreational and professional users of GNSS products benefit from the level of accuracy and precision achievable in everyday applications such as surveying, mapping, GIS, construction, machine control, tourism and many others. These kinds of applications are well known and well covered in several textbooks and trade publications. This appendix only summarizes a few of the applications in the areas of geohazards monitoring, infrastructure monitoring, and environmental monitoring to illustrate the role of the science of GNSS geodesy in solving some of the problems and challenges faced by humanity.

C.1 Earth and Land Deformation in Millimeters

Figure C.1 shows the horizontal velocity field of the continental USA, Alaska, and a few locations in Mexico and Canada with respect to the International Terrestrial Reference Frame (ITRF) (source: ngs.noaa.gov/CORS). Figure C.2 is an example of time series from a single station in the CORS network. The velocity vectors as shown in Fig. C.1 are derived from such datasets. Figures C.3 and C.4 show examples of regional land subsidence and earthquake geohazards that are captured and monitored through long-term continuous GNSS measurements.

C. A. Ogaja, *Introduction to GNSS Geodesy*,
https://doi.org/10.1007/978-3-030-91821-7

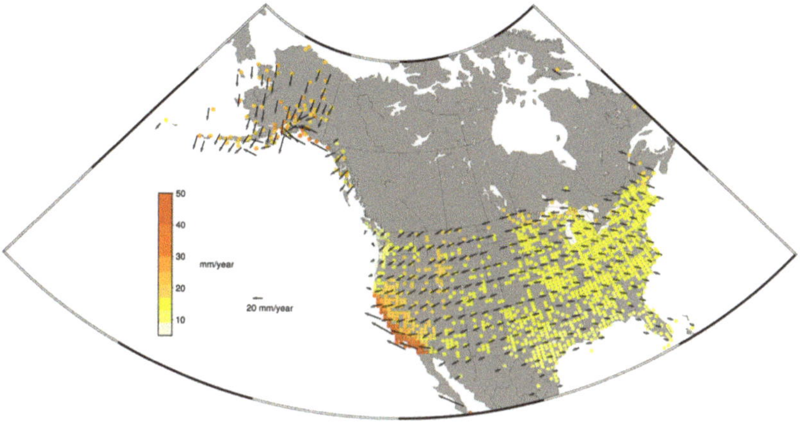

Fig. C.1 Horizontal velocity field derived from CORS relative to ITRF2014 showing the North American plate tectonics. According to the color scale, some places move at few millimeters per year, while in some places the movement is as fast as 40–50 mm per year (source: ngs.noaa.gov/CORS)

Fig. C.2 Geodetic time series of a CORS station UPSA by Nevada Geodetic Laboratory (geodesy.unr.edu/NGLStationPages/stations/UPSA.sta)

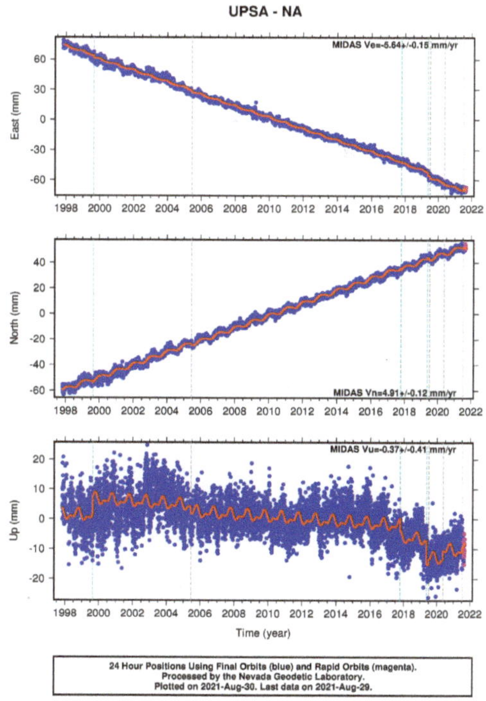

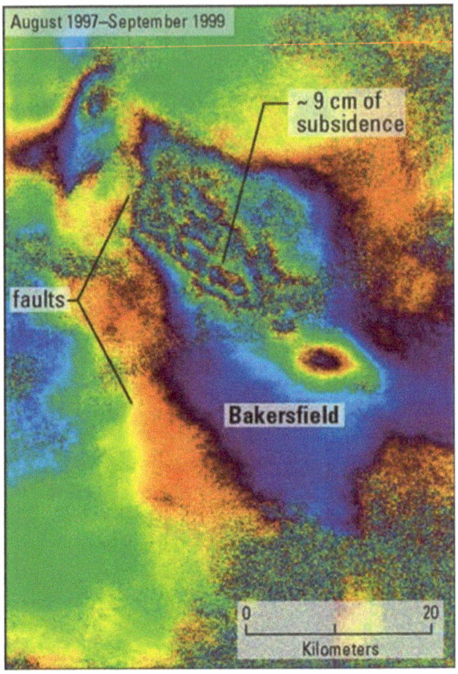

Fig. C.3 Ground subsidence in Bakersfield, California (Source: USGS)

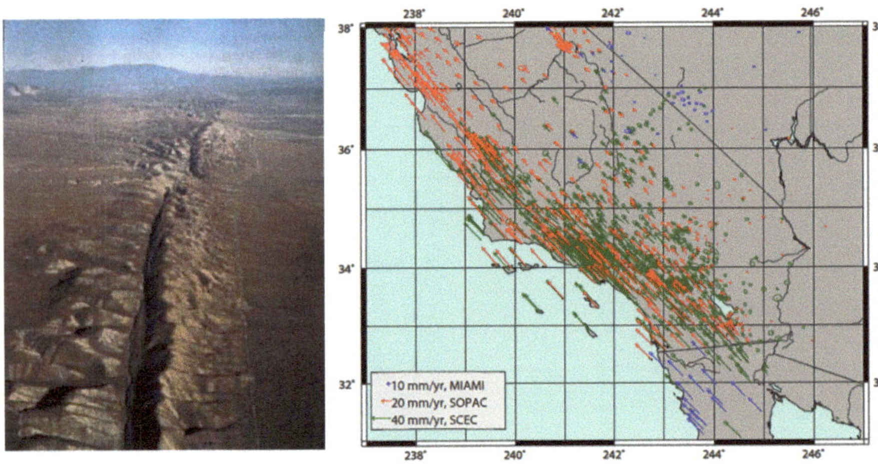

Fig. C.4 Plate tectonics at San Andreas Fault, California (Source: USGS)

C.2 GNSS Instrumentation of Tall Buildings and Other Structures in Hurricane/Typhoon-Prone Areas

Numerous studies have shown that GNSS instrumentation of tall buildings and other infrastructures such as long-span suspension bridges records wind-induced

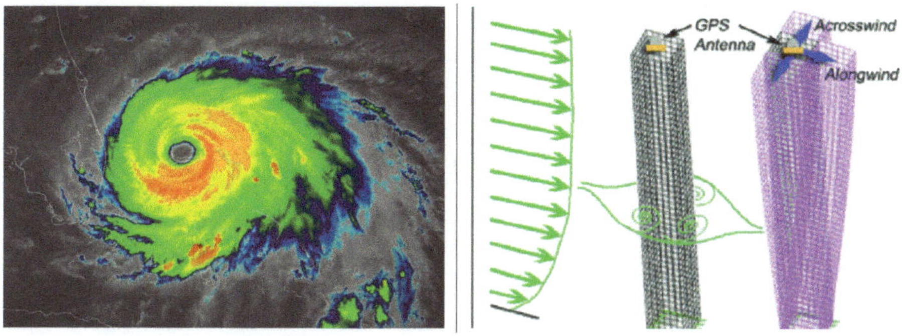

Fig. C.5 GPS/GNSS instrumentation of tall buildings and infrastructure such as bridges are capable of recording precise data capturing structural integrity during hurricanes and typhoons (see, e.g., [3, 6, 7])

response during hurricanes and typhoons. GNSS data recorded during such events are used to analyze the structural performance for improved safety and design (see, e.g., [3, 4, 6–8]). Pioneering work and field trials on tall buildings, towers and bridges, complemented with GNSS receiver technology developments and innovative research, have led to operational systems in various places worldwide (see, e.g., in [8]; Fig. C.5).

C.3 GNSS Interferometric Reflectometry (GNSS-IR)

GNSS-IR, a measurement technique based on indirect GNSS signals, is used to study environmental phenomena. The Cyclone Global Navigation Satellite System (CYGNSS) shown in Fig. C.6 is an example (see details in [10]). CYGNSS, a space-based GNSS-IR system, measures the distortion of GPS signals scattered from the ocean surface to determine ocean surface roughness and wind speed. This technique is used for improved hurricane forecasting by better understanding the interactions between the ocean and the air near the core of a storm. Other GNSS-IR applications include, for example, soil moisture monitoring with ground-based GNSS CORS stations [1], observing sea level variations [2, 5], and other similar applications [9].

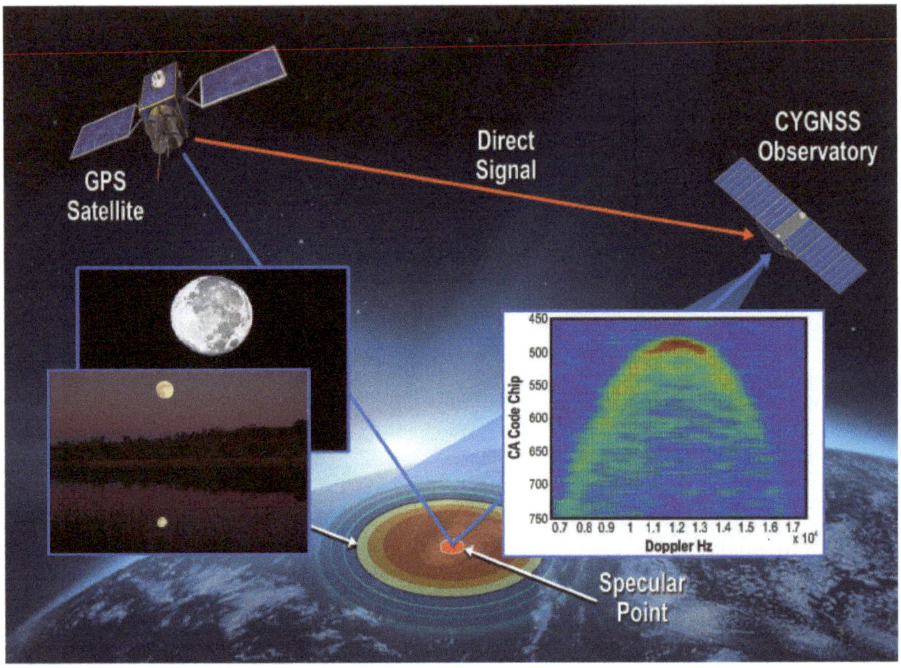

Fig. C.6 Cyclone Global Navigation Satellite System (CYGNSS). Image used by permission. Courtesy of Prof. Christopher S. Ruf, University of Michigan [10]

References

1. Chew, C. C., Small, E. E., & Larson, K. M. (2016). An algorithm for soil moisture estimation using GPS interferometric reflectometry for bare and vegetated soil. *GPS Solutions, 20*(3), 525–537. https://doi.org/10.1007/s10291-015-0462-4
2. Holden, L., & Larson, K. M. (2021). Ten years of Lake Taupo surface height estimates using the GNSS interferometric reflectometry. *Journal of Geodesy, 95*(74). https://doi.org/10.1007/s00190-021-01523-7
3. Kijewski-Correa, T., & Kareem, A. (2003, September). The height of precision. *GPS World, 14*(9), 20–34.
4. Kijewski-Correa, T., & Kareem, A. (2004). The height of precision: New perspectives in structural monitoring. In *Proceedings of Engineering, Construction, and Operations in Challenging Environments: Earth & Space* (pp. 195–201). https://doi.org/10.1061/40722(153)28
5. Larson, K. M., Lay, T., Yamazaki, Y., Cheung, K. F., Ye, L., Williams, S. D. P., & Davis, J. L. (2021). Dynamic sea level variation from GNSS: 2020 Shumagin earthquake tsunami resonance and Hurricane Laura. *Geophysical Research Letters, 148*(4). https://doi.org/10.1029/2020GL091378
6. Ogaja, C. (2011). *Applied GPS for engineers and project managers*. Reston: ASCE Press. https://doi.org/10.1061/9780784411506
7. Ogaja, C., Li, X., & Rizos, C. (2007). Advances in structural monitoring with Global Positioning System technology: 1997–2006. *Journal of Applied Geodesy, 1*, 171–179. https://doi.org/10.1515/jag.2007.019

8. Ogaja, C., Rizos, C., Wang, J., & Brownjohn, J. M. W. (2001, June 5–8). A dynamic GPS system for on-line structural monitoring. In *International Symposium on Kinematic Systems in Geodesy, Geomatics & Navigation (KIS2001), Banff, Canada* (pp. 290–297).
9. Roesler, C. J., & Larson, K. M. (2018). Software tools for GNSS interferometric reflectometry. *GPS Solutions, 22*, 80. https://doi.org/10.1007/s10291-018-0744-8
10. Ruf, C., et al. (2017, January 22–26). The NASA CYGNSS satellite constellation for tropical cyclone observations. In *AMS Annual Meeting 21st Conference on Integrated Observing and Assimilation Systems for the Atmosphere, Oceans and Land Surface Seattle, WA.*

Appendix D
Linear Model for Phase Observable

Let us consider the case of single differenced GNSS phase observables of between-station differences. We use the operator Δ to indicate a difference between stations (receivers).

Differencing between receivers is in effect estimating the baseline vector between the two receivers, rather than the two positions.

The formulation of the linear model in Eq. 7.7 of Chap. 7 can be explained as follows.

Let us look at the linear model for a case of 2 stations, 4 satellites. We will consider station A to be fixed and estimate station B. This means there will be four (4) single differenced (SD) observations every epoch, and seven (7) unknown parameters to be estimated (3 coordinate vectors and 4 SD ambiguities).[1] Hence at least two epochs of data are needed for a least squares estimation.

As an example, consider the row of the design matrix $\underline{\mathbf{A}}$ (Eq. 7.7 of Chap. 7) corresponding to the second satellite (2) at an epoch t of an observable and the column vector of unknowns as follows:

$$\begin{pmatrix} & & & \vdots & & & \\ \frac{\partial\Delta\Phi_{AB}^{(2)}(t)}{\partial X_B} & \frac{\partial\Delta\Phi_{AB}^{(2)}(t)}{\partial Y_B} & \frac{\partial\Delta\Phi_{AB}^{(2)}(t)}{\partial Z_B} & \frac{\partial\Delta\Phi_{AB}^{(2)}(t)}{\partial\Delta N_{AB}^{(1)}} & \frac{\partial\Delta\Phi_{AB}^{(2)}(t)}{\partial\Delta N_{AB}^{(2)}} & \frac{\partial\Delta\Phi_{AB}^{(2)}(t)}{\partial\Delta N_{AB}^{(3)}} & \frac{\partial\Delta\Phi_{AB}^{(2)}(t)}{\partial\Delta N_{AB}^{(4)}} \\ & & & \vdots & & & \\ & & & \vdots & & & \\ & & & \vdots & & & \end{pmatrix} \begin{pmatrix} \delta x_B \\ \delta y_B \\ \delta z_B \\ \Delta N_{AB}^{(1)} \\ \Delta N_{AB}^{(2)} \\ \Delta N_{AB}^{(3)} \\ \Delta N_{AB}^{(4)} \end{pmatrix} \tag{D.1}$$

[1] Assume some terms, e.g., clock offsets and biases, are known, ignored, or modeled.

C. A. Ogaja, *Introduction to GNSS Geodesy*,
https://doi.org/10.1007/978-3-030-91821-7

This leads to the corresponding linear model:[2]

$$\begin{pmatrix} \vdots \\ \lambda\Delta\Phi_{AB}^{(2)}(t) - \rho_{B(0)}^{(2)}(t) + \rho_{A}^{(2)}(t) \\ \vdots \\ \vdots \\ \vdots \end{pmatrix} = \begin{pmatrix} & & \vdots & & & & \\ \frac{\partial\rho_{AB}^{(2)}(t)}{\partial X_B} & \frac{\partial\rho_{AB}^{(2)}(t)}{\partial Y_B} & \frac{\partial\rho_{AB}^{(2)}(t)}{\partial Z_B} & 0 & \lambda & 0 & 0 \\ & & \vdots & & & & \\ & & \vdots & & & & \\ & & \vdots & & & & \end{pmatrix} \begin{pmatrix} \delta x_B \\ \delta y_B \\ \delta z_B \\ \Delta N_{AB}^{(1)} \\ \Delta N_{AB}^{(2)} \\ \Delta N_{AB}^{(3)} \\ \Delta N_{AB}^{(4)} \end{pmatrix} \tag{D.2}$$

$$\begin{pmatrix} \vdots \\ \lambda\Delta\Phi_{AB}^{(2)}(t) - \rho_{B(0)}^{(2)}(t) + \rho_{A}^{(2)}(t) \\ \vdots \\ \vdots \\ \vdots \end{pmatrix}$$

$$= \begin{pmatrix} & & \vdots & & & & \\ \frac{X_{B(0)}-X^{(2)}(t)}{\rho_{B(0)}^{(2)}(t)} & \frac{Y_{B(0)}-Y^{(2)}(t)}{\rho_{B(0)}^{(2)}(t)} & \frac{Z_{B(0)}-Z^{(2)}(t)}{\rho_{B(0)}^{(2)}(t)} & 0 & \lambda & 0 & 0 \\ & & \vdots & & & & \\ & & \vdots & & & & \\ & & \vdots & & & & \end{pmatrix} \begin{pmatrix} \delta x_B \\ \delta y_B \\ \delta z_B \\ \Delta N_{AB}^{(1)} \\ \Delta N_{AB}^{(2)} \\ \Delta N_{AB}^{(3)} \\ \Delta N_{AB}^{(4)} \end{pmatrix} \tag{D.3}$$

Once all the rows for all epochs are formulated as shown above, a least squares estimation process leads to the estimated coordinates of station B and the corresponding error estimates:

$$X_B = X_{B(0)} + \delta x_B;\ \sigma_{X_B};\ Y_B = Y_{B(0)} + \delta y_B;\ \sigma_{Y_B};\ Z_B = Z_{B(0)} + \delta z_B;\ \sigma_{Z_B} \tag{D.4}$$

Components of the baseline vector between stations A and B can then be derived:

$$\begin{aligned} dX_{AB} &= X_B - X_A; & \sigma_{dX_{AB}} &= \sqrt{(\sigma_{X_B})^2 + (\sigma_{X_A})^2} \\ dY_{AB} &= Y_B - Y_A; & \sigma_{dY_{AB}} &= \sqrt{(\sigma_{Y_B})^2 + (\sigma_{Y_A})^2} \\ dZ_{AB} &= Z_B - Z_A; & \sigma_{dZ_{AB}} &= \sqrt{(\sigma_{Z_B})^2 + (\sigma_{Z_A})^2} \end{aligned} \tag{D.5}$$

[2] Assumes a short baseline and ignores relative receiver clock offsets and biases.

Index

C. A. Ogaja, *Introduction to GNSS Geodesy*,
https://doi.org/10.1007/978-3-030-91821-7

MIX
Papier | Fördert gute Waldnutzung
FSC® C083411

Kolibri 360 GmbH
Ferdinand-Jühlke-Straße 7
99095 Erfurt, Deutschland
produktsicherheit@kolibri360.de